Edexcel International GCSE Physics Simplified

About the author

Kaleem Akbar was born on the outskirts of Glasgow, Scotland in 1980. He graduated with a B.Sc. Honours degree in Optoelectronics and Laser Engineering from Heriot Watt University in 2002 and went on to do an M.Sc. at the University of St Andrews, before completing his PGDE (Post Graduate Diploma in Education) in Physics at Strathclyde University in Glasgow. He taught in Scotland before moving out to the Middle East and has taught Physics at IGCSE, AS and A2 level since September 2006. He wrote this book in response to his students' thirst for the essential details to achieve their full potential.

Acknowledgements

The author would like to thank the following professionals who have helped to make this book possible:

Lynda Anderson-Coe for editing the Physics content
Philippa Sage for copy editing
Mark Venables for examiner's review
Jane Roth, **Geoff Amor** and **Kate Blackham** for proofreading
Craig Walton for creating and editing the illustrations
Shahad Al Qattan for front cover design
Craig Walton for front cover editing.

A special thanks to my wife **Toni Reid** for her unflappable support and to my friends **Alistair Rae** and **Emily Anne Fair** for their counsel.

Thanks are also due to the countless students who have provided their invaluable feedback and recommendations for improving my original notes over the years.

Edexcel International GCSE Physics Simplified

Kaleem Akbar

www.igcsephysics.com

Edexcel International GCSE Physics Simplified
© Kaleem Akbar 2016

All rights reserved

No part of this book may be reproduced in any form by photocopying or any electronic or mechanical means, including information storage or retrieval systems, without permission in writing from both the copyright owner and the publisher of the book.

British Library Cataloguing in Publication Data

A catalogue record for this publication is available from the British Library.

Endorsement Statement

In order to ensure that this resource offers high-quality support for the associated Pearson qualification, it has been through a review process by the awarding body. This process confirms that this resource fully covers the teaching and learning content of the specification or part of a specification at which it is aimed. It also confirms that it demonstrates an appropriate balance between the development of subject skills, knowledge and understanding, in addition to preparation for assessment. Endorsement does not cover any guidance on assessment activities or processes (e.g. practice questions or advice on how to answer assessment questions), included in the resource nor does it prescribe any particular approach to the teaching or delivery of a related course. While the publishers have made every attempt to ensure that advice on the qualification and its assessment is accurate, the official specification and associated assessment guidance materials are the only authoritative source of information and should always be referred to for definitive guidance. Pearson examiners have not contributed to any sections in this resource relevant to examination papers for which they have responsibility. Examiners will not use endorsed resources as a source of material for any assessment set by Pearson. Endorsement of a resource does not mean that the resource is required to achieve this Pearson qualification, nor does it mean that it is the only suitable material available to support the qualification, and any resource lists produced by the awarding body shall include this and other appropriate resources.

ISBN 978-178456-433-9
First published 2016 by Fast-Print Publishing

Important information for students

Pearson Edexcel International GCSE Physics (4PH1) specification

This book is designed to cover all of the learning outcomes for the latest Pearson Edexcel International GCSE Physics (4PH1) specification for examinations from 2019 onwards. You should be aware that the specification can vary slightly from year to year.

This book contains material offering students the opportunity to obtain an International GCSE in either Physics or the Physics content of Science (Double Award 4SD0).

Features in the book

Symbols

❑ Double Award material
○ Additional material for Separate Science Award

Students preparing for the Separate Science Award in Physics should study both the Double Award material and the additional material.

N.B. is used at the beginning of some sentences. It is an abbreviation for a Latin expression and means 'note well'. It indicates that the sentence is particularly important.

Remember: is also used at the beginning of some sentences to remind you of information you have previously been given.

Examples

The book contains examples of the types of calculations you may have to perform. All questions and answers have been written by the author.

The answers are given to two or three significant figures (sig. figs).
For an in-depth explanation of significant figures see pages 270-71.

Top Tip

Top Tip boxes contain useful advice and guidance that will help you to work to the best of your abilities.

Note

Note boxes summarise or highlight some important ideas.

Units

The SI system of units is the world's most widely used system of measurement. SI is an abbreviation for le Système International d'Unités.

The system has seven base units, including the kilogram (kg), metre (m) and second (s).

There are other derived units, including the metre per second (m/s), as well as recognised multipliers such as mega (M), kilo (k), centi (c) and milli (m).

All of the units used in this book are part of the SI system.

How to best use this book

- You can show that you have learned each statement/explanation with a tick in the box on pages viii and ix or next to each square or circle bullet in the main text.
- A column has been left on the outside of each page so that you can make additional notes.
- Highlighter pens can also be used to emphasise important statements throughout this book.
- At the back of the book you will find additional support material, including a list of all the equations that you have to know expressed in different formats, including the triangular format. Simply cover up the quantity you want to calculate and the triangle will show whether the other quantities have to be multiplied or divided.
- There is also a glossary of terms at the back of the book. This is a brief dictionary that will help you to understand any words or phrases you're not sure of.

Exam structure

The assessment for the Edexcel International GCSE in Physics is linear. Students sit Paper 1 (2 hours) and Paper 2 (1 hour 15 minutes).

Paper 1 assesses material marked with this symbol ❏ only.

Paper 2 assesses material marked with symbols ❏ and ◯.

Students studying for the Edexcel International GCSE in Science (Double Award) sit Paper 1 in Physics as well as Paper 1 in Chemistry and Paper 1 in Biology.

The papers contain multiple-choice questions, short-answer structured questions and some questions requiring longer answers. You will be assessed on your knowledge and understanding and on your ability to apply that knowledge and understanding.

You may be required to describe experiments and analyse and evaluate data, including drawing graphs and performing calculations.

Good examination habits

- Make sure you are fully equipped for your examination; ensure you have pens, pencils, a pencil sharpener, a ruler, a rubber/eraser, a calculator, a protractor and a pair of compasses.
- Draw all diagrams in pencil.
- Draw graphs in pencil; ensure the drawn line is not thicker than the grid lines on the graph paper. Make sure that any best fit line is drawn in one sweeping movement using a ruler for a straight line and free hand for a curve. When drawing a curve, ensure that your wrist is on the inside of the curve (if necessary, you will have to rotate the question paper). Plot Xs rather than •s on your graph. See pages 272-75 for further advice on drawing graphs.
- Always show fully your working for numerical questions. It is far better to score some marks than none at all. Remember to include units for all calculated quantities in your answers to numerical questions.
- Ensure that you re-read the questions and not just your answers when you have completed your exam. You may find you have a correct answer but for an entirely different question.
- Working through past examination questions is an important part of your preparation.

Revision checklist

■ shaded sub-topics contain Separate Science material from the specification

Tick box once revised. (The more times you revise each topic the better.)

Topic	Sub-topic	1	2	3	4
Forces and motion	1a Units				
	1b Movement and position				
	1c Forces, movement, shape and momentum				
Electricity	2a Units				
	2b Mains electricity				
	2c Energy and voltage in circuits				
	2d Electric charge				
Waves	3a Units				
	3b Properties of waves				
	3c The electromagnetic spectrum				
	3d Light and sound				
Energy resources and energy transfers	4a Units				
	4b Energy transfers				
	4c Work and power				
	4d Energy resources and electricity generation				
Solids, liquids and gases	5a Units				
	5b Density and pressure				
	5c Change of state				
	5d Ideal gas molecules				
Magnetism and electro-magnetism	6a Units				
	6b Magnetism				
	6c Electromagnetism				
	6d Electromagnetic induction				

			Tick box once revised. (The more times you revise each topic the better.)			
			1	2	3	4
Radioactivity and particles	7a	Units				
	7b	Radioactivity				
	7c	Fission and fusion				

Astrophysics	8a	Units				
	8b	Motion in the universe				
	8c	Stellar evolution				
	8d	Cosmology				

Contents

Edexcel International GCSE Physics	iv
Important information for students	v
Revision checklist	viii

Section 1   Forces and motion
1a	Units	2
1b	Movement and position	6
1c	Forces, movement, shape and momentum	20

Section 2   Electricity
2a	Units	53
2b	Mains electricity	53
2c	Energy and voltage in circuits	62
2d	Electric charge	84

Section 3   Waves
3a	Units	91
3b	Properties of waves	91
3c	The electromagnetic spectrum	98
3d	Light and sound	102

Section 4   Energy resources and energy transfers
4a	Units	123
4b	Energy transfers	123
4c	Work and power	135
4d	Energy resources and electricity generation	147

Section 5   Solids, liquids and gases
5a	Units	154
5b	Density and pressure	154
5c	Change of state	166
5d	Ideal gas molecules	179

Section 6   Magnetism and electromagnetism
6a	Units	191
6b	Magnetism	191
6c	Electromagnetism	197
6d	Electromagnetic induction	207

Section 7   Radioactivity and particles
7a	Units	217
7b	Radioactivity	217
7c	Fission and fusion	232

Section 8   Astrophysics
8a	Units	236
8b	Motion in the universe	236
8c	Stellar evolution	249
8d	Cosmology	255

Additional support material
1	Physical quantities and units	262
2	How to use equations effectively	265
3	Equations	265
4	Working with numbers	270
5	Graphs	272

Glossary	276
Glossary of examination terminology	281
Index	282

"Everything should be made as simple as possible, but not simpler."

Attributed to Albert Einstein 1879 – 1955

Section 1 Forces and motion

Section 1a Units

❑ In this section you will come across the following units:
- the **metre** (m) is the unit of length
- the **kilogram** (kg) is the unit of mass
- the **second** (s) is the unit of time
- the **metre per second** (m/s) is the unit of speed or velocity
- the **metre per second squared** (m/s^2) is the unit of acceleration
- the **newton** (N) is the unit of force
- the **newton per kilogram** (N/kg) is the unit of gravitational field strength

○ And also:
- the **newton metre** (Nm) is the unit of moment (turning effect of a force)
- the **kilogram metre per second** (kgm/s) is the unit of momentum.

❑ When making measurements, physicists use various instruments, such as rulers to measure length, balances to measure mass and stopwatches to measure time.

Length

❑ A ruler is used to measure length. When using a ruler, be careful to avoid **parallax** error.

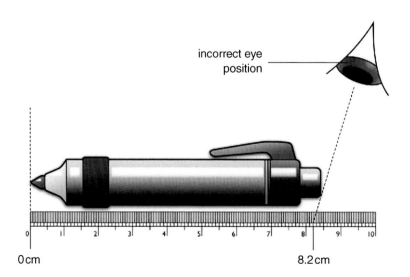

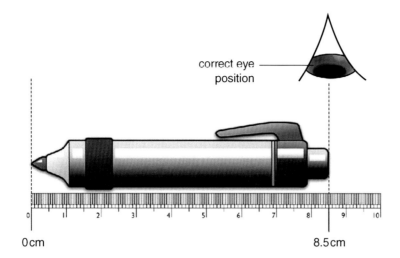

0cm   8.5cm

Note

Parallax causes an object to appear to move relative to the ruler scale, so that you measure it to be shorter or longer depending on how you view it. An object must be viewed at right angles and as close to the scale as possible to measure its length correctly.

❑ There are many units of length but scientists use a system called the SI (Système International) system. The SI unit of length is the **metre** (m).

❑ To get a more **accurate** value for measurements, repeat the measuring process several times and calculate a mean (average) value of your readings.

❑ It is difficult to measure very small lengths. If you have, say, 50 sheets of paper, it is possible to measure the thickness of all 50 sheets and then calculate the mean thickness of one sheet by dividing by 50.

Similarly, if you measure the height of a column of 20 coins you can calculate the mean thickness of one coin by dividing the height of the column by 20.

N.B. This does not tell you the actual thickness of any one coin.

Mass

- The **mass** of an object is a measure of the amount of matter in the object and can be measured using an electronic balance.

- In general we often talk about 'weighing' an object when we want to find its mass. We shall see later that mass and weight are different (see pages 25-27).

electronic balance

Time

- **Time** can be measured with an analogue stopwatch or clock. These are mechanical devices, as they operate from a coiled spring, which operates a series of gear wheels as it unwinds.

- Measuring a very small amount of time, such as the period of one pendulum swing, can be very difficult, especially if the arc of the swing is small. The time it would take you to react – your **reaction time** – would affect the measurement.

- By measuring the time taken for many (10 or 20) swings (oscillations), the period can be calculated by dividing the total time taken by the number of swings.

$$\text{period} = \frac{\text{total time}}{\text{number of swings}}$$

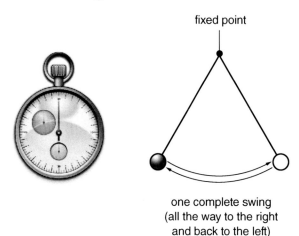

one complete swing
(all the way to the right
and back to the left)

- A digital stopwatch can measure time to 0.01 of a second and so will give a **more precise** reading than using a mechanical stopwatch with a moving second hand. It could still be **inaccurate** because of the reaction time of the person using it.

- The digital stopwatch below shows a time of 3 minutes and 43.00 seconds.

- Digital timers are often used in connection with electronic circuits and can be switched on and off by sensors. For example, a light gate is often used in timing experiments. A light gate has a beam of light from a source which travels to a detector. If an object passes between the source and the detector, a timer measures how long the beam is cut for. Such timers can measure time intervals with high accuracy as well as high precision.

Top Tip

 Repeating measurements, checking any **anomalous** results and removing them if they are wrong, and calculating a **mean** value are all likely to improve the accuracy of an experiment.

Section 1b Movement and position

Distance–time graphs

❑ A simple graph can be plotted showing the distance moved by an object in a given time.

N.B. See pages 272-75 for more information about drawing graphs.

Stationary
(Not moving, at rest)

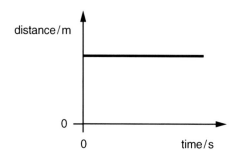

Constant speed
(No acceleration)

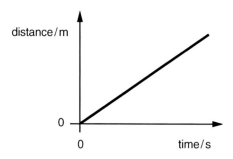

Increasing speed
(Acceleration)

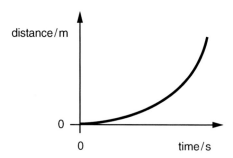

Decreasing speed
(Deceleration)

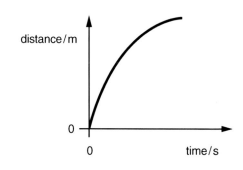

Variable speed
(Speeding up, then constant speed, slowing down and then stopping – coming to rest)

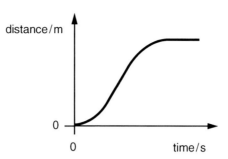

❏ The speed can be calculated from the **gradient** (slope) of a distance–time graph. The steeper the line, the faster the object is travelling.

Note

In mathematics the gradient is given by:

$m = \dfrac{\Delta y}{\Delta x}$ m = gradient
          Δy = change in y
          Δx = change in x

For a distance–time graph:

$\text{gradient} = \dfrac{\text{distance moved}}{\text{time taken}} = \text{speed}$

❏ **Speed** is a measure of how fast something is moving or the distance moved per unit of time.

❏ **Average speed** is the speed measured over a period of time.

❏ **Instantaneous speed** is the speed measured at an instant in time.

❏ **Speed**, or **velocity**, has the symbol v, or sometimes u, and its unit is the **metre per second** (m/s).

❏ **Distance** has the symbol s, or sometimes d, and its unit is the **metre** (m).

❏ **Time** has the symbol t, and its unit is the **second** (s).

Note

Speed and **velocity** are both terms used to describe how fast an object moves, although we will see later that **velocity** is also related to the direction in which the object moves.

- These quantities are related by the equation:

$$\text{average speed} = \frac{\text{distance moved}}{\text{time taken}}$$

$$v = \frac{s}{t}$$

v = average speed (m/s)
s = distance moved (m)
t = time taken (s)

Velocity–time graphs

- A graph can also be plotted showing how the velocity of an object varies with time.

- If the velocity increases with time, the object is said to be accelerating.

- If the velocity decreases with time, the object has negative acceleration, which can be referred to as deceleration.

- The shape of a velocity–time graph allows us to describe the motion of the object.

Constant velocity
(Steady, uniform velocity)
(No acceleration)
(At rest if $v = 0$)

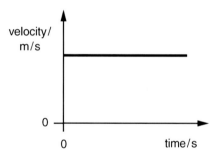

Constant positive acceleration
(Speeding up)
(Increasing velocity)

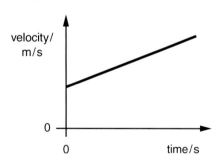

Constant negative acceleration
(Constant deceleration)
(Slowing down)
(Decreasing velocity)

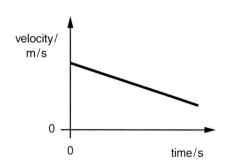

Changing acceleration
(Increasing acceleration)
(Gradient of curve increases)
(Velocity increases at an increasing rate)

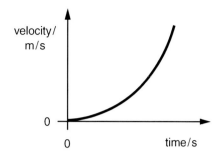

Changing acceleration
(Decreasing acceleration)
(Gradient of curve decreases)
(Velocity increases at a decreasing rate)

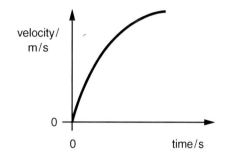

Changing deceleration
(Decreasing deceleration)
(Gradient of curve decreases)
(Velocity decreases at a decreasing rate)

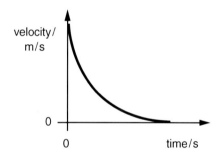

Varying velocity from zero to a maximum value and back to zero
A increasing acceleration
B constant acceleration
C constant velocity
D constant deceleration
E decreasing deceleration

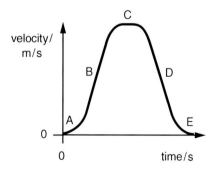

☐ Acceleration, velocity and time are related by the following equation.

$$\text{acceleration} = \frac{\text{change in velocity}}{\text{time taken}}$$

$$a = \frac{(v - u)}{t}$$

a = acceleration (m/s²)
v = final velocity (m/s)
u = initial velocity (m/s)
t = time taken (s)

☐ The information from a **velocity–time graph** can be used to calculate various values.

- The **acceleration** is the **gradient** of the graph. Acceleration has the symbol **a**, and it is the rate of change of velocity (how quickly an object becomes faster or slower). Its unit is the **metre per second squared** (m/s²).
- The **distance moved** is equal to the **area under** the velocity–time graph.
  N.B. The area is the total area down to the time axis.
- The **maximum acceleration** can be found by choosing the part of the graph with the steepest gradient and then calculating the gradient.

Note

In mathematics the gradient is given by:

$$m = \frac{\Delta y}{\Delta x}$$

m = gradient
Δy = change in y
Δx = change in x

For a velocity–time graph the gradient m gives the acceleration. A negative gradient on a velocity–time graph shows that the object is slowing down or decelerating.

Example 1b (i)

The velocity–time graph below represents a motorbike going on a very short journey from the rider's house (A) to a local store (F). Use the graph to:

(a) describe the motion at the various stages of the journey AB, BC, CD, DE and EF

(b) calculate the acceleration at the various stages of the journey, giving all answers to two significant figures

(c) state the maximum velocity during the journey

(d) calculate the total distance moved.

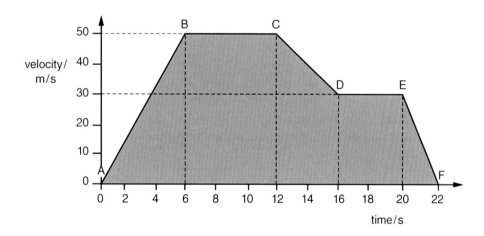

Answer

(a) Stage AB: constant acceleration

Stage BC: constant velocity

Stage CD: constant deceleration

Stage DE: constant velocity

Stage EF: constant deceleration

(b) AB: $a = \dfrac{(v - u)}{t} = \dfrac{(50 - 0)}{6.0} = 8.3 \, \text{m/s}^2$

CD: $a = \dfrac{(v - u)}{t} = \dfrac{(30 - 50)}{4.0} = -5.0 \, \text{m/s}^2$

EF: $a = \dfrac{(v - u)}{t} = \dfrac{(0 - 30)}{2.0} = -15 \, \text{m/s}^2$

For BC and DE $a = 0$

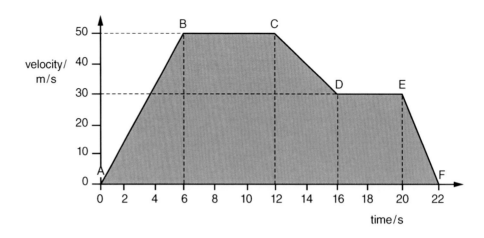

(c) Maximum velocity = 50 m/s

(d) Distance moved = area under the graph.

AB: $\frac{1}{2}(6.0 \times 50) = 150$ m

BC: $(6.0 \times 50) = 300$ m

CD: $\frac{1}{2}[(50 - 30) \times 4.0] + (4.0 \times 30) = 40 + 120 = 160$ m

DE: $(4.0 \times 30) = 120$ m

EF: $\frac{1}{2}(30 \times 2.0) = 30$ m

Total distance = 150 + 300 + 160 + 120 + 30 = 760 m

ALWAYS REMEMBER TO STATE THE UNIT FOR CALCULATED QUANTITIES.

Example 1b (ii)
A car travels at 20 m/s for 1 minute and 10 seconds.

(i) State the equation linking average speed, distance and time.
(ii) Calculate the distance moved.

Answer

(i) $\text{average speed} = \dfrac{\text{distance moved}}{\text{time taken}}$ or $v = \dfrac{s}{t}$

(ii) **Step 1** List all the information in symbol form and change into appropriate and consistent SI units if required.

$v = 20\,\text{m/s}$
$t = 1 \text{ minute and } 10 \text{ seconds} = 70\,\text{s}$
$s = ?$

Step 2 Rearrange the equation.

$v = \dfrac{s}{t} \implies s = v \times t$

Step 3 Calculate the answer by putting the numbers into the equation.

$s = 20 \times 70 = 1400\,\text{m}$

ALWAYS REMEMBER TO STATE THE UNIT FOR CALCULATED QUANTITIES.

Note

When substituting numbers into equations and performing calculations, it is important that the units of the variables are consistent. For example, if the distance *s* is in m and the time *t* is in s then the speed *v* will be in m/s.

Investigating the motion of an object

- By measuring the time taken for an object to travel a given distance, you can calculate the speed.

- If you release a toy car at the top of a ramp which has been marked off into 10 cm sections, you and your friends could use stopwatches to measure the time taken to reach the 10 cm mark, the 20 cm mark and so on. However, the times will be short and your reaction times will make the measurements inaccurate.

- The following experiment describes how to investigate the motion of an object using a ticker tape timer. A ticker tape timer is a very simple device which can be used to measure small intervals of time accurately.

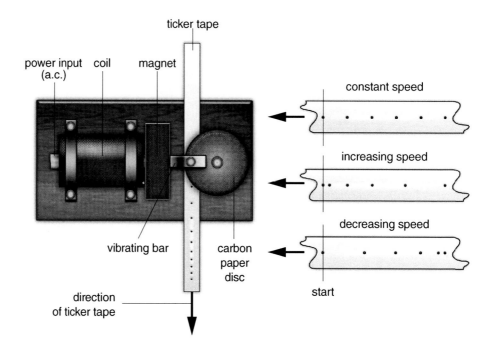

- The vibrating bar on a ticker tape timer moves up and down 50 times a second. If a piece of tape is pulled through as shown by the arrow, a dot, or tick, is recorded on the tape every 1/50th of a second.

- The space between the ticks is the distance travelled by the tape in 1/50th of a second. If the tape is pulled through at constant speed, the ticks are evenly spaced. If the speed increases, the space between the ticks increases, and, if the speed decreases, the space between the ticks decreases.

- [ ] The distance travelled in 5 ticks (i.e. the distance between tick 1 and tick 6) can be measured with a ruler, and because there are 50 ticks made every second, this is the distance travelled in 1/10th of a second.

- [ ] The average speed for that 5-tick period can be calculated using the equation:

    $v = \dfrac{s}{t}$ where s is the length of the 5-tick section of tape in cm and t is 0.1 s.

    Remember: This is an average value of speed in that 5-tick period; the instantaneous speed may be increasing or decreasing.

Method

1. Attach the tape to the object whose motion you want to investigate, e.g. a toy car.

2. Switch on the timer and release the car down the ramp.

3. Cut the tape into 5 (or 10) tick lengths and stick them vertically in order to create a graph as shown in the diagram below.

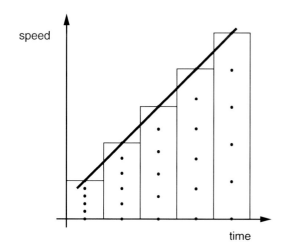

4. The graph is a speed–time graph. In this case it shows that the distance travelled in successive 5 ticks increases, and so the speed increases. In other words, the car accelerates.

    It is a uniform acceleration because a line drawn through the centres of the top of the columns is a straight line.

Equations of motion

❑ The acceleration equation can be rearranged as:

$$v = u + (a \times t)$$

v = final velocity (m/s)
u = initial velocity (m/s)
a = acceleration (m/s²)
t = time taken (s)

❑ When the **initial velocity** u is zero, the equation can be more simply expressed as:

$$v = a \times t$$

Example 1b (iii)

A toy car accelerates from rest at 10 cm/s² for 18 s.

(i) State the equation linking acceleration, velocity and time.
(ii) Calculate the final velocity.

Answer

(i) $\text{acceleration} = \dfrac{\text{change in velocity}}{\text{time taken}}$ or $a = \dfrac{(v - u)}{t}$

(ii) **Step 1** List all the information in symbol form and change into appropriate and consistent SI units if required.

$u = 0$ because the toy car starts from rest
$a = 10 \text{ cm/s}^2 = 0.1 \text{ m/s}^2$
$t = 18 \text{ s}$
$v = ?$

Step 2 Rearrange the equation.

$a = \dfrac{(v - u)}{t} \implies v = u + (a \times t)$

Step 3 Calculate the answer by putting the numbers into the equation

$v = 0 + (0.1 \times 18) = 1.8 \text{ m/s}$

ALWAYS REMEMBER TO STATE THE UNIT FOR CALCULATED QUANTITIES.

1b Movement and position

- We have seen equations that relate speed or velocity, distance, time and acceleration. These equations only apply if the acceleration is constant or uniform.

- For an object accelerating uniformly, the distance moved can be found using the following equation:

    distance moved = average speed × time taken

    $$s = \left(\frac{u + v}{2}\right) \times t$$

- This can be rearranged to express time taken in terms of v, u and s:

    $$t = \frac{2s}{(v + u)}$$

- Sometimes you will be expected to do a calculation without knowing the time taken, starting with this equation you already know:

    $v = u + (a \times t)$

    Substituting for t and simplifying gives:

    $$\boxed{v^2 = u^2 + (2 \times a \times s)}$$

    v = final speed (m/s)
    u = initial speed (m/s)
    a = acceleration (m/s^2)
    s = distance moved (m)

Note

These are the equations you should know and be able to use:

average speed = $\dfrac{\text{distance moved}}{\text{time taken}}$ $v = \dfrac{s}{t}$

acceleration = $\dfrac{\text{change in velocity}}{\text{time taken}}$ $a = \dfrac{(v - u)}{t}$

and $v^2 = u^2 + (2 \times a \times s)$

Note

Always consider the number of significant figures you give in your answers to calculations. In the following example the information given (values of u, v and s) is to 2 significant figures, so you should not express your answer to more than 2 sig. figs (see pages 270-71).

Example 1b (iv)

An aircraft has to reach a speed of 45 m/s before it can lift off. Calculate the acceleration required so the aircraft (plane) can take off before it reaches the end of the runway if the runway is 150 m long. The equation linking final and initial speed, acceleration and distance moved is given below.

$$v^2 = u^2 + (2 \times a \times s)$$

Answer

Step 1 List all the information in symbol form and change into appropriate and consistent SI units if required.

$v = 45$ m/s
$u = 0$ m/s
$s = 150$ m
$a = ?$

Step 2 Rearrange the equation.

$$v^2 = u^2 + (2 \times a \times s) \implies a = \frac{(v^2 - u^2)}{2s}$$

Step 3 Calculate the answer by putting the numbers into the equation.

$$a = \frac{(45^2 - 0^2)}{2 \times 150} = 6.75 = 6.8 \text{ m/s}^2 \text{ (to 2 sig. figs)}$$

This is the minimum acceleration required.

ALWAYS REMEMBER TO STATE THE UNIT FOR CALCULATED QUANTITIES.

Example 1b (v)

A quad bike decelerates uniformly from 3.5 m/s to 1.0 m/s in 10 s.

(i) State the equation linking average speed, time taken and distance moved.
(ii) Calculate the distance moved by the bike.

Answer

(i) average speed = $\dfrac{\text{distance moved}}{\text{time taken}}$ or $v = \dfrac{s}{t}$

N.B. It is important to understand that the symbol *v* can be used to express speed or velocity. In this example, *v* is used for both average speed and final speed.

1b Movement and position

(ii) **Step 1** List all the information in symbol form and change into appropriate and consistent SI units if required.

initial speed $u = 3.5\,\text{m/s}$
final speed $v = 1.0\,\text{m/s}$
so average speed $= \left(\dfrac{u+v}{2}\right)$
$t = 10\,\text{s}$
$s = ?$

Step 2 Calculate the answer by putting the numbers into the equation.

$$s = \left(\dfrac{u+v}{2}\right) \times t$$

$$s = \left(\dfrac{3.5 + 1.0}{2}\right) \times 10 = 22.5 = 23\,\text{m} \quad \text{(to 2 sig. figs)}$$

Example 1b (vi)

A sketch of the velocity–time graph for the journey in the previous example is given below. Use the information from the graph to calculate the distance moved.

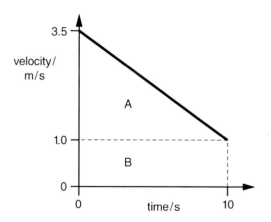

Answer

Distance moved equals the area under the graph.
Remember: the area must go down to the axis.

Area A $= \dfrac{1}{2}(3.5 - 1.0) \times 10 = 12.5\,\text{m}$

Area B $= 10 \times 1.0 = 10\,\text{m}$

Area A + Area B $= 12.5 + 10 = 22.5 = 23\,\text{m}$ (to 2 sig. figs)

ALWAYS REMEMBER TO STATE THE UNIT FOR CALCULATED QUANTITIES.

Section 1c Forces, movement, shape and momentum

- ❏ A **force** is a pull, push, twist, stretch or squeeze. We can observe the effects of forces but cannot see the actual forces themselves.

- ❏ Forces can:
    - change the velocity (speed in a particular direction) of an object
    - change the direction of movement of an object
    - change the shape of an object
    - change the size of an object.

- ❏ Some forces are contact forces, when one object is in contact with another; for example, when a rubber band is stretched, someone or something is pulling it.

- ❏ Other forces act at a distance; for example the gravitational force that exists between the Sun and a planet orbiting it, the magnetic force between poles of two magnets or between one magnet and a piece of iron, or the electrostatic force between electric charges.

- ❏ If the forces acting on an object in opposite directions are equal, the forces are said to be **balanced**. The sum or **resultant** of the forces is zero in such cases. (In the following diagram all the values of **F** are the same.)

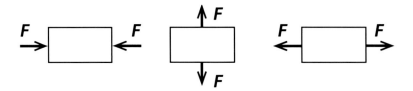

- ❏ When there are **no forces** acting, or the forces acting on an object are balanced, the object may be **stationary** or moving at a **constant velocity**.

- ❏ Force has **magnitude** (size) and **direction**, so it is a **vector** quantity. A quantity which has **magnitude** only is called a **scalar** quantity.

- ❑ When an object is moving at a constant speed in a straight line, there is **no resultant force**. The forces on it are equal and in opposite directions.

- ❑ If the forces on an object are not balanced, there will be a resultant force that will cause the object to speed up (accelerate), slow down or change direction, depending on the direction of the force. The greater the force, the greater the acceleration.

- ❑ The resultant force is the **overall unbalanced force**. It is the **vector sum** of all the forces on an object (i.e. taking into consideration the direction of the forces).

Adding and subtracting to calculate resultant forces

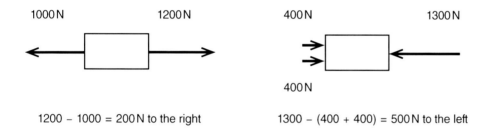

1200 − 1000 = 200 N to the right   1300 − (400 + 400) = 500 N to the left

- ❑ Force is a vector, so it always has a direction.

- ❑ If forces are acting along a line but in opposite directions, they should be subtracted. If forces are acting in the same direction, they should be added.

- ❑ The same principles are used for forces acting vertically upwards and downwards, e.g. the forces on a falling object.

Scalar and vector quantities

- ❑ **Speed** is a **scalar** quantity because it has magnitude (size) only and can be described as the distance moved per unit time.

- ❑ **Velocity** is a **vector** quantity because it has magnitude and direction. It can be described as the speed in a particular direction.

- ❑ It follows that if a force changes the direction of a moving object, it changes its velocity, even though the speed might stay the same. An example of this is a planet moving in a circular orbit.

❏ The table below gives examples of scalar and vector quantities:

Scalar	Vector
speed	velocity
time	acceleration
distance	force
energy	weight
mass	momentum

Newton's Laws of Motion

Sir Isaac Newton formulated laws that describe the motion of an object when it has forces acting on it.

❏ **Newton's 1st Law** states that an object remains **at rest** or **moves** at a **steady speed** in a **straight line** unless acted on by a **resultant** or **unbalanced force**.

❏ **Newton's 2nd Law** states that an object **accelerates** in the direction of a **resultant** or **unbalanced force**.

❏ **Resultant force** has the symbol *F*, and its unit is the **newton** (N).

❏ **Mass** has the symbol *m*, and is a measure of the amount of matter in an object. Its unit is the **kilogram** (kg).

❏ **Acceleration** has the symbol *a*, and it is the rate of change of velocity (how quickly an object becomes faster or slower). Its unit is the **metre per second squared** (m/s²).

❏ Resultant force, mass and acceleration are related by the following equation:

force = mass × acceleration

$$F = m \times a$$

F = resultant force (N)
m = mass (kg)
a = acceleration (m/s²)

- ❑ When the mass of an object is constant and the **resultant force** applied to the object **increases**, the **acceleration increases**.

- ❑ The greater the mass of an object, the smaller the acceleration for a given force.

Example 1c (i)

A jet ski accelerates at 2.0 m/s² and has a resultant force of 540 N in the direction of motion.

(i) State the equation linking force, mass and acceleration.
(ii) Calculate the mass of the jet ski and rider.

Answer

(i) Force = mass × acceleration   or   **F = m × a**

(ii) **Step 1** List all the information in symbol form and change into appropriate and consistent SI units if required.

$$F = 540\,N$$
$$a = 2.0\,m/s^2$$
$$m = ?$$

Step 2 Rearrange the equation.

$$F = m \times a \quad \Rightarrow \quad m = \frac{F}{a}$$

Step 3 Calculate the answer by putting the numbers into the equation.

$$m = \frac{540}{2} = 270\,kg$$

ALWAYS REMEMBER TO STATE THE UNIT FOR CALCULATED QUANTITIES.

Newton's 3rd Law

- **Newton's 3rd Law** states that when body A exerts a force on body B, body B exerts an **equal and opposite** force on body A.

- In other words for every **action** (force) there is an equal and opposite **reaction** (force). Always state the object that each force is acting on. For example, when you swim you push backwards on the water with your arms. The water then pushes forwards on you, making you move.

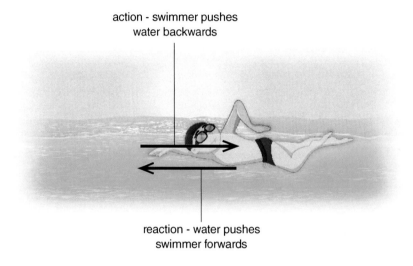

- To understand Newton's 3rd Law we need to consider some more practical examples.

- When you sit on a chair, you push down on it with a contact force. Newton's 3rd Law states that the chair pushes up on you with an equal and opposite contact force. If it did not, you would fall to the floor!

- When a space rocket is started in deep space, the fuel ignites and combines with oxygen to produce exhaust gases that are directed out of the bottom of the rocket. The rocket pushing the gas backwards is the action, which is accompanied by an equal and opposite reaction of the gas pushing the rocket forwards.

- When a car is started, the wheels start to move and the **friction force** between the tyres and the road surface pushes backwards on the road. The road exerts an equal and opposite force to push the car forwards. In icy conditions or on an oily surface the friction is much less and it is difficult to move the car forwards.

Mass and weight

- **Weight** has the symbol **W**, and its unit is the **newton** (N).

- The weight of an object is defined as the **force due to gravity** acting on an object's mass.
  Remember: Weight is a force. **Never** refer to weight as just 'gravity'; it is the **force due to gravity.**

- **Mass** has the symbol **m**, and its unit is the **kilogram** (kg).

- Mass is a property that resists change in motion. The greater the mass, the greater the force needed to set an object in motion, to stop it or to change its direction.

- The mass of an object will not change if the strength of the gravitational field changes. The weight, however, will change if the strength of the gravitational field changes.

- The relationship between weight, mass and **gravitational field strength** can be expressed as:

  weight = mass × gravitational field strength

  $$W = m \times g$$

  W = weight (N)
  m = mass (kg)
  g = gravitational field strength (N/kg)

- On Earth the gravitational field strength, **g**, is 10 N/kg.

- As **g** is a constant near the surface of Earth, the weight of an object is proportional to its mass. If mass doubles, weight doubles.

- Force, and hence weight, can be measured using a newton meter, also known as a spring balance.

- The **mass** of an object is a measure of the amount of matter in the object. To measure mass we can use an electronic balance.

- Mass and weight are terms that are interchangeable in everyday language but in physics they are different things.

❑ We 'weigh' ourselves on bathroom scales and yet the figure we read is given in kilograms, the unit of mass. In fact, most bathroom scales do measure weight but, because *g* is a constant near the surface of Earth and mass is directly proportional to weight, the scale can be marked in kilograms. If you took your bathroom scales to the Moon you would get a different reading (see page 27).

Example 1c (ii)

On Earth a man has a mass of 85 kg.

(i) State the equation linking weight, mass and the gravitational field strength.

(ii) Calculate his weight on Earth if *g*, the gravitational field strength, is 10 N/kg.

Answer

(i) Weight = mass × gravitational field strength   or   *W* = *m* × *g*

(ii) **Step 1**   List all the information in symbol form and change into appropriate and consistent SI units if required.

m = 85 kg
g = 10 N/kg
W = ?

Step 2   Calculate the answer by putting the numbers into the equation.

W = 85 × 10 = 850 N

ALWAYS REMEMBER TO STATE THE UNIT FOR CALCULATED QUANTITIES.

> **Top Tip**
>
>
> The **mass** of an object will remain the **same** whether it is on Venus, Mars or Earth or anywhere else. This is because the **amount of matter** in the object stays the **same**. The **weight changes** because the **gravitational field strength** is different on different planets.

❑ The diagram below shows how the mass remains constant whether an astronaut is on the Moon or on Earth. Only the weight changes due to the different gravitational field strengths.

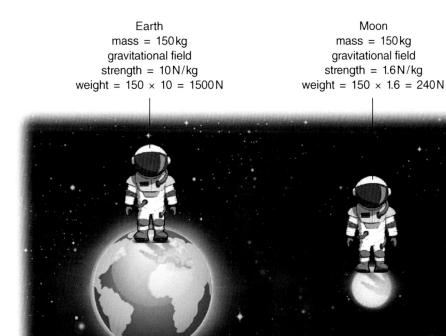

Earth
mass = 150 kg
gravitational field
strength = 10 N/kg
weight = 150 × 10 = 1500 N

Moon
mass = 150 kg
gravitational field
strength = 1.6 N/kg
weight = 150 × 1.6 = 240 N

Centre of gravity

○ The **centre of gravity** (c. of g.) of an object is the point through which all the weight appears to be acting.

○ This is a useful simplification because we can assume that the **force of gravity** only acts at a **single point**.

○ This means a single arrow can represent the weight **W** of an object, as shown in the diagram of the tower opposite.

centre of gravity

W

- Although the tower is leaning, it does not topple because the line of action of its weight falls within the base (see 'turning effect', pages 46-47).

- The centre of gravity for objects with a regular shape is in the centre. Drawing dashed lines from various points that exactly cut the shape into two equal parts can easily help find the centre of gravity as shown below. A minimum of two lines is required.

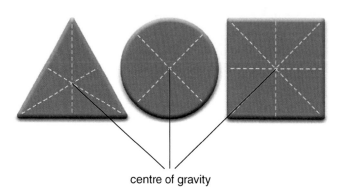

centre of gravity

Top Tip

The centre of gravity is the point through which all the weight appears to act. A ruler will balance perfectly when its centre of gravity is placed directly over a pivot. Otherwise it will tip over.

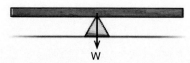 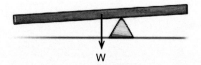

Note

Racing cars travel at very high speeds and need to be very stable. They have a small height, which gives them **a low centre of gravity**. A racing car is also wide so that if it starts to tip, the line of action of its weight stays within the base and it rights itself.

In general:
wide and short = stable (low centre of gravity)
tall and thin = less stable (high centre of gravity)

Friction

- ☐ The force of **friction opposes the motion** of an object.

- ☐ **Drag** is the term used for the frictional opposing force when objects travel through fluids such as air, water and oil. **Air resistance** is a form of drag.

- ☐ Drag **increases** the **faster** an object travels through a fluid.

- ☐ Friction causes **heating**, e.g. when one material is pushed across the surface of another.

- ☐ Friction can be useful in certain situations and not useful in others.

- ☐ Friction is **not useful** in the following situations and can be **decreased** by:
    - skiers putting wax on their skis to make them smooth
    - making objects **streamlined**, allowing them to travel faster by cutting through the air or water
    - oiling engines to allow the parts to move easily.

- ☐ Friction is **useful** in the following situations and can be **increased** by:
    - pressing the brake pedals in a car, slowing it down quickly
    - a skydiver opening a parachute allowing him/her to slow down quickly due to increased air resistance.

Example 1c (iii)

A Formula One car has a mass of 640 kg (including the driver). When the engine exerts a driving force of 29 000 N, the opposing frictional force is 810 N.

(i) State the equation linking resultant force, mass and acceleration.

(ii) Calculate the acceleration.

Answer

(i) Force = mass × acceleration   or   **F = m × a**

(ii) **Step 1**   Draw a diagram showing the forces acting on the object and their directions.

Step 2   Calculate the **resultant** (**unbalanced**) force.

$$F = 29\,000 - 810 = 28\,190\,\text{N}$$

Step 3   List all the information in symbol form and change into appropriate and consistent SI units if required.

$F = 28\,190\,\text{N}$
$m = 640\,\text{kg}$
$a = ?$

Step 4   Rearrange the equation.

$$F = m \times a \quad \Rightarrow \quad a = \frac{F}{m}$$

Step 5 Calculate the answer by putting the numbers into the equation.

$$a = \frac{28190}{640} = 44.05 = 44\,\text{m/s}^2 \quad \text{(to 2 sig. figs)}$$

ALWAYS REMEMBER TO STATE THE UNIT FOR CALCULATED QUANTITIES.

Falling objects

❑ When an object is dropped from a height and falls to the ground, the force acting on it that causes it to fall is the force due to gravity.

❑ Any object falling under gravity near the Earth's surface will initially accelerate at 10 m/s². This is known as the **acceleration of free fall**. If there is no **air resistance** then the object will continue to speed up by 10 m/s every second.

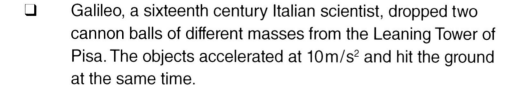

❑ The acceleration of free fall (10 m/s²) is numerically equal to the gravitational field strength (10 N/kg) and has the symbol **g**, the same as that for gravitational field strength. The two terms are sometimes used interchangeably, although their definitions are different and they have different units.

❑ Galileo, a sixteenth century Italian scientist, dropped two cannon balls of different masses from the Leaning Tower of Pisa. The objects accelerated at 10 m/s² and hit the ground at the same time.

But if one of the objects dropped had a parachute attached, it would be more affected by air resistance and its acceleration would be reduced. The two objects would not then reach the ground at the same time.

❑ Any object **thrown upwards** decelerates at 10 m/s² (ignoring air resistance).

- There are two forces acting on a falling object:
  - the **weight** of the object which is a downwards force caused by the Earth's gravitational field
  - **air resistance**, a frictional force which acts in the opposite direction to the movement. On a falling object the air resistance acts vertically upwards.

- Weight is a constant force depending only on mass and the gravitational field strength **g**.

- Air resistance opposes the movement and **increases with the speed of the object**, i.e. the faster the object is moving, the greater the air resistance.

- The **resultant** of the two forces, weight and air resistance, determines the motion of the falling object.

- At first the object is moving slowly and the air resistance is low. As the object falls it accelerates at $10\,m/s^2$ and the air resistance increases; the resultant downward force, and hence the acceleration, decreases.

- When the resultant force is zero, the velocity of the falling object becomes constant. This velocity is called **terminal velocity**.

- When a parachutist jumps from a small aircraft, he begins to fall to the ground as a result of the force due to gravity (his weight). At first his acceleration is $10\,m/s^2$.

  The graph opposite shows how his velocity varies with time during his descent.
  - As he accelerates, air resistance opposes his motion and decreases his acceleration. At A the gradient of the graph is decreasing. The faster he travels, the greater the air resistance. In other words, the air resistance increases as he falls.
  - Eventually he will reach terminal velocity at B when the air resistance is equal to the force due to gravity (weight). Because the two forces are in opposite directions, there is no resultant (net) force (see page 20).
  - This terminal velocity is too high for him to land safely
  - When he opens his parachute, the air resistance increases greatly because of the very large surface area of the parachute.

- The air resistance is greater than the force due to gravity. He continues to fall but decelerates rapidly, as shown at point C.
- As he slows down, the air resistance decreases until once again he reaches terminal velocity at D, when the air resistance equals the force due to gravity. This terminal velocity is much lower and will allow him to land safely.
- When he lands his velocity becomes zero.

❑ **Remember:** Weight is also known as the force due to gravity.

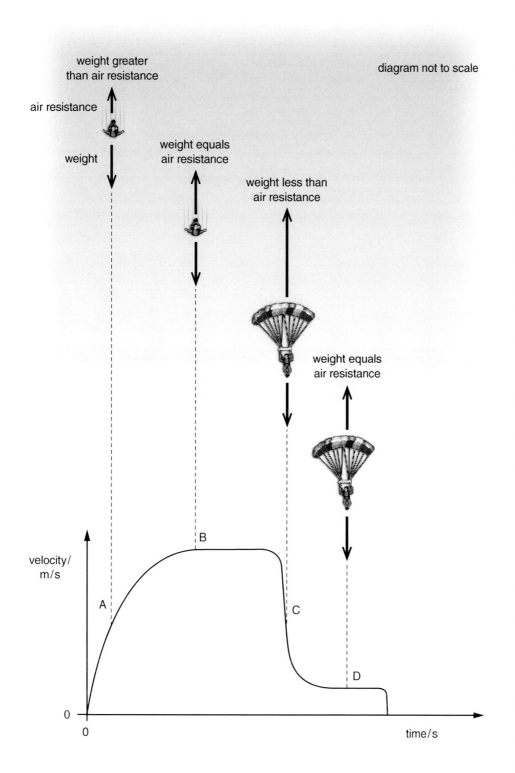

The stopping distance of a car

❑ The **stopping distance** is the distance a car travels after the driver applies the brakes. It is made up of the sum of the thinking distance and the braking distance.

stopping distance = thinking distance + braking distance

❑ The **thinking time** is the time it takes a driver to react and start to apply the brakes. During this time the car will continue to travel at the same speed. The distance it travels before the driver reacts is called the **thinking distance**.

❑ There are many factors that can affect the driver's reaction time, including:
- how tired he or she is
- whether he or she is under the influence of alcohol or drugs
- whether he or she is unwell
- whether he or she is distracted by loud music or a noisy conversation or using a cell (mobile) phone.

❑ The driver then applies the brakes. The distance the car travels before coming to a stop is called the **braking distance**.

❑ There are many factors that affect the braking distance, including:
- the speed of the car when the driver applies the brakes
- whether the brakes or tyres are worn
- what the weather conditions are like
- if the car is heavily laden, i.e. the mass of the car
- the type of road surface
- whether there is oil or water on the road surface
- the aerodynamics of the car.

❑ The graph opposite shows some typical stopping distances.

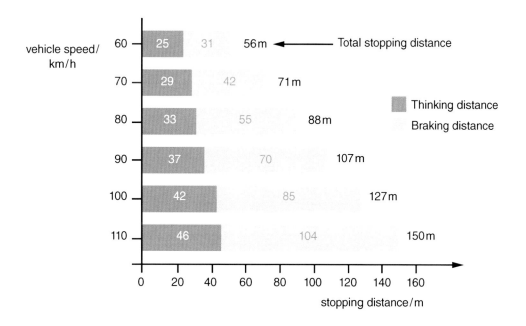

- Note that the stopping distance of the vehicle shown in the graph increases with speed, i.e. the faster the car is going, the greater the distance it travels before coming to a stop.

- The thinking distance increases with speed at a constant rate. This is because it is related to the reaction time of the driver. If the car is moving faster, it will travel further in the same length of time.

- The braking distance increases with speed at an increasing rate.

Example 1c (iv)

(i) State the equation relating average speed, distance and time.
(ii) Using the graph above calculate the driver's thinking time when travelling at 60 km/h.

Answer

(i) average speed = $\dfrac{\text{distance moved}}{\text{time taken}}$ or $v = \dfrac{s}{t}$

(ii) **Step 1** List all the information in symbol form and change into appropriate and consistent SI units if required.

$v = 60 \text{ km/h} = \dfrac{60\,000}{60 \times 60} = 16.7 \text{ m/s}$

$t = ?$

$s = 25 \text{ m}$ (the thinking distance, from the graph)

Step 2 Rearrange the equation.

$$v = \frac{s}{t} \quad \Rightarrow \quad t = \frac{s}{v}$$

Step 3 Calculate the answer by putting the numbers into the equation.

$$t = \frac{25}{16.7} = 1.499 = 1.5\,\text{s} \quad (\text{to 2 sig. figs})$$

ALWAYS REMEMBER TO STATE THE UNIT FOR CALCULATED QUANTITIES.

Momentum

- The **momentum** of a moving object is defined as the product of the mass and the velocity.

- These quantities are related by the equation:

momentum = mass × velocity

$$p = m \times v$$

p = momentum (kg m/s)
m = mass (kg)
v = velocity (m/s)

- Momentum has the symbol p, and its unit is the **kilogram metre per second (kg m/s)**.
  Remember: Mass has the symbol m, and its unit is the **kilogram** (kg), and velocity has the symbol v, and its unit is the **metre per second** (m/s).

- Momentum is a vector quantity; it has magnitude (size) and direction. The momentum of an object will change if either the mass or the velocity changes, and this includes changing direction.

- Any mass that is moving has momentum.
  - An oil tanker (large boat) has a large momentum even when it is moving slowly because it has a very large mass.
  - A speedboat has a small momentum in comparison to the oil tanker because, although it can move much faster, it has a much smaller mass.

Example 1c (v)

(i) State the equation linking mass, velocity and momentum.
(ii) Calculate the momentum of a car with a mass of 1000 kg travelling due east at 25 m/s, and give its direction.

Answer

(i) momentum = mass × velocity or $p = m \times v$

(ii) **Step 1** List all the information in symbol form and change into appropriate and consistent SI units if required.

m = 1000 kg
v = 25 m/s
p = ?

Step 2 Calculate the answer by putting the numbers into the equation.

p = 1000 × 25 = 25000 kg m/s

Momentum is a vector quantity and so the answer should read: The momentum is 25000 kg m/s due east.

ALWAYS REMEMBER TO STATE THE UNIT FOR CALCULATED QUANTITIES.

○ If a similar car were travelling at the same speed due west we would say its momentum was −25000 kg m/s because it is travelling in the opposite direction.

Newton's 2nd Law and momentum

○ You have previously learned (see page 22) that according to Newton's 2nd Law:

$F = m \times a$

F = force (N)
m = mass (kg)
a = acceleration (m/s^2)

- The acceleration can be calculated by using:

$$a = \frac{(v - u)}{t}$$

  a = acceleration (m/s²)
  v = final velocity (m/s)
  u = initial velocity (m/s)
  t = time taken (s)

- The acceleration equation can be substituted into $F = m \times a$ to give:

$$F = m \times \frac{(v - u)}{t} = \frac{(m \times v) - (m \times u)}{t} = \frac{(mv - mu)}{t}$$

- Since generally $p = m \times v$ then, in this case, $m \times v$ is the final momentum and $m \times u$ is the initial momentum. Therefore

$$\text{force} = \frac{\text{change in momentum}}{\text{time taken}}$$

- Newton's 2nd Law of motion can therefore be worded as:
  The rate of change of momentum of an object is equal to the resultant force acting on it.
  If the mass of the object is constant, this is a different way of saying $F = m \times a$.

- The same force acting for the same time produces a greater effect on the motion of a small mass (e.g. a speedboat) than on the motion of a large mass (e.g. an oil tanker).

Safety features

- When a moving object comes to a sudden stop, there is a change in momentum in a very short time and therefore a large force.

  This is because of the equation:

$$\text{force} = \frac{\text{change in momentum}}{\text{time taken}}$$

- It follows that if the time taken for the **same change of momentum** is increased, the force is smaller. This is used to great effect in many safety devices, particularly in cars.

- If a car crashes, the people inside experience a very large force because the time for the car and for them to stop is so short. This can cause serious injuries.

- A seat belt prevents people from being thrown around in a car, but it also has enough 'give' to allow the passenger to move forward slightly. This means that the **time taken for the same change in momentum is increased**, so the force is decreased. An airbag works in a similar way; the passenger keeps travelling a short distance after a crash, increasing the time for the change in momentum.

- Many modern cars have crumple zones which are designed to fold and buckle on impact. This is another way of increasing the time taken for a car to stop in an accident and decreasing the force on the people inside the car.

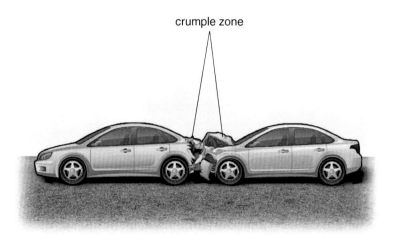

- A cycle helmet works in a similar way. The inside of a helmet is soft and designed to move. If the cyclist comes to an abrupt stop and his head hits the ground, the helmet will increase the time taken for the skull to come to rest, so decreasing the rate of momentum change and the force on the skull.

Example 1c (vi)

A model train of mass 4.0kg travels along a straight track. Its velocity increases from 2.0m/s to 9.0m/s in 5.0s.

(i) State the equation linking force, change in momentum and time.

(ii) Calculate the average force acting on the train.

Answer

(i) $\text{force} = \dfrac{\text{change in momentum}}{\text{time taken}}$ or $F = \dfrac{(mv - mu)}{t}$

(ii) **Step 1** List all the information in symbol form and change into appropriate and consistent SI units if required.

$m = 4.0\,\text{kg}$
$u = 2.0\,\text{m/s}$
$v = 9.0\,\text{m/s}$
$t = 5.0\,\text{s}$
$F = ?$

Step 2 Calculate the answer by putting the numbers into the equation.

$$F = \dfrac{(4.0 \times 9.0) - (4.0 \times 2.0)}{5.0} = 5.6\,\text{N}$$

ALWAYS REMEMBER TO STATE THE UNIT FOR CALCULATED QUANTITIES.

Conservation of momentum

○ The **principle of conservation of momentum** states that the total momentum of objects before they collide is equal to their total momentum after the collision provided no external forces act on the objects.

$$\text{total momentum before collision} = \text{total momentum after collision}$$

This principle allows us to predict how an object will move after a collision. A golf club striking a ball, a meteor crashing into a planet and two cars crashing into one another are all examples of collisions.

Imagine a toy lorry of mass m_1 travelling at a velocity of u_1 collides with a toy car of mass m_2 travelling at a velocity of u_2. After the collision, their velocities are v_1 and v_2, respectively, as shown below. We can write:

$$(m_1 \times u_1) + (m_2 \times u_2) = (m_1 \times v_1) + (m_2 \times v_2)$$

Before collision

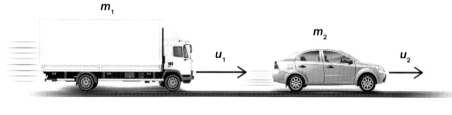

After collision

Note

It is common practice to denote an object moving from left to right as having a positive velocity and momentum, and an object moving from right to left as having negative velocity and momentum. If a calculation gives you a negative value for velocity, it means the object is moving in the opposite direction to an object with a positive value for velocity.

Example 1c (vii)

A ball A of mass 1.0 kg moving to the right at a velocity of 4.0 m/s collides with another ball B of mass 1.0 kg, which is stationary.

(i) State the equation linking momentum, mass and velocity.
(ii) Calculate the total momentum before the collision.
(iii) Calculate the velocity of ball B after the collision if ball A is then stationary.

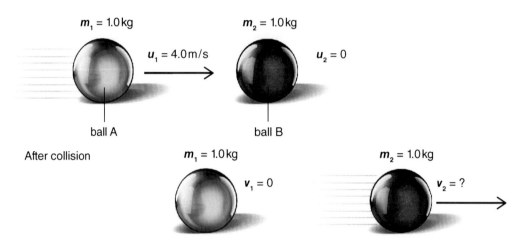

Answer

(i) momentum = mass × velocity   or $p = m \times v$

(ii) **Step 1** List all the information in symbol form and change into appropriate and consistent SI units if required.

Before collision:
$m_1 = 1.0$ kg
$u_1 = +4.0$ m/s (i.e. moving to the right)
$m_2 = 1.0$ kg
$u_2 = 0$
$p = ?$

Step 2 Calculate the answer by putting the numbers into the equation.

$$p = m \times v$$
for m_1, $p_1 = 1.0 \times 4.0 = 4.0$ kg m/s
for m_2, $p_2 = 1.0 \times 0 = 0$ kg m/s
total momentum = 4.0 + 0 = 4.0 kg m/s

ALWAYS REMEMBER TO STATE THE UNIT FOR CALCULATED QUANTITIES.

(iii) **Step 1** List all the information in symbol form and change into appropriate and consistent SI units if required.

After collision:
$p = 4.0 \text{ kgm/s}$ because momentum is conserved
$m_1 = 1.0 \text{ kg}$
$v_1 = 0 \text{ m/s}$
$m_2 = 1.0 \text{ kg}$
$v_2 = ?$

Step 2 Calculate the answer by putting the numbers into the equation.

$$p = m \times v$$

for m_1, $p_1 = 1.0 \times 0$
for m_2, $p_2 = 1.0 \times v$

But we know momentum is conserved so

$$p_1 + p_2 = 4.0 \text{ kgm/s}$$
$$(1.0 \times 0) + (1.0 \times v) = 4.0$$
$$0 + v = 4.0$$
$$v = 4.0 \text{ m/s}$$

Because v is positive, ball B moves to the right at 4.0 m/s.

ALWAYS REMEMBER TO STATE THE UNIT FOR CALCULATED QUANTITIES.

Hooke's Law

❏ Hooke's Law states that the **extension x** of an elastic object such as a spring or metal wire is directly proportional to the **force F** applied. This means that if the force is doubled, the extension is doubled.

$$F \propto x$$

❏ The following experiment investigates Hooke's Law using a spring.

Method

1. Assemble the apparatus as shown in diagram A, ensuring the ruler is vertical. (In a laboratory environment a clamp stand, metre rule, spring, mass hanger and slotted masses are needed.)

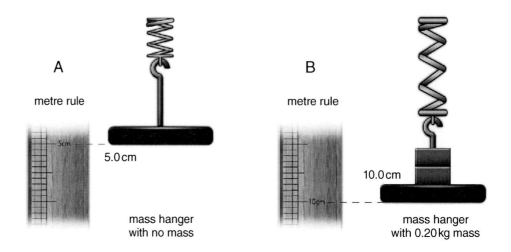

2. The experiment involves extending a spring by adding weights to it. If too many weights are added the spring might snap or permanently deform. For safety reasons, therefore, you should wear safety spectacles and make sure the weights are not likely to fall onto your feet.

3. Note and record the reading on the scale of the rule next to the bottom of the mass hanger without adding any masses. In diagram A this is 5.0 cm.

4. Add one slotted mass (the weight or load of a 100 g mass is 1.0 N) to the hanger and measure the extension on the ruler. The extension is calculated by

   extension = new length − original length

5. Repeat step 4, adding one mass at a time and recording the corresponding extension reading. (Diagram B shows two 100 g masses added. In this case the extension would equal the length of the spring with the 2.0 N load minus the length of the spring with no load which is 10.0 − 5.0 = 5.0 cm.)

6. Prepare a table of your results for load (calculated from the mass readings) and extension.

7. Plot a graph of extension against load. Your graph should be a straight line passing through the origin. This shows that the extension is directly proportional to the load (weight or force).

☐ The spring should return to its original length if you remove the masses. This means it shows **elastic behaviour**. Elasticity is the ability of a material to return to its original (undeformed) shape.

☐ If you continued to add masses your graph would look like the one below. The **limit of proportionality** is shown and is where the straight line ends.

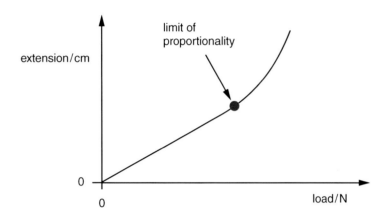

Note

If any two quantities are **directly proportional**, when they are plotted against each other on a graph, the graph has two characteristics.
- It passes through the **origin**.
- It is a **straight line**.

For Hooke's Law, if the load (force) is doubled then the extension doubles up to the limit of proportionality. In other words, the gradient of the extension–load graph is constant up until this point.

☐ **Force** has the symbol **F** and its unit is the **newton** (N).

☐ **Extension** of a spring has the symbol **x** and is measured in mm, cm or m.

☐ These quantities are related by the equation:

$$F = k \times x$$

F = force (N)
k = spring constant (N/m)
x = extension (m)

- ❏ The **spring constant k** can be calculated by rearranging the above formula:

  $$F = k \times x \quad \Rightarrow \quad k = \frac{F}{x}$$

- ❏ The **limit of proportionality** is the point beyond which the spring extension **will not be proportional** to the load. Up to this limit the extension increases by a set amount for every newton of force applied. Above this limit the increase in extension per newton will be greater.

- ❏ The limit of proportionality is an important point.
  - If the spring is stretched beyond this point, it will no longer obey Hooke's Law.
  - If the spring is stretched only up to this point, the spring will obey Hooke's Law.

- ❏ If a similar experiment was carried out using a rubber band instead of a spring it would be shown that, although the rubber band extends, it does not obey Hooke's Law.

Turning effect

- ○ Sometimes a force causes an object to rotate or turn. A **moment** is a measure of the turning effect of a force.

- ○ The turning effect depends on two things:
  - the **magnitude** of the force applied
  - the **perpendicular distance** of the force from the **pivot**.

- ○ A pivot is the point around which rotational movement takes place.

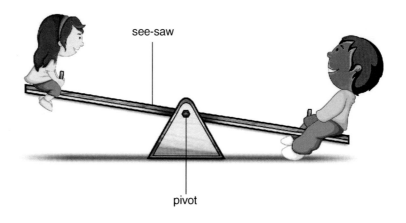

- **Moment** may be given the symbol **M**, and its unit is the **newton metre** (Nm).
  Remember: The unit of force is the newton (N) and the unit of distance is the metre (m).

- These quantities are related by the equation:

  moment = force × perpendicular distance from the pivot

  $$M = F \times d$$

  M = moment (Nm)
  F = force (N)
  d = perpendicular distance from the pivot (m)

- The moment is greater if either **F** or **d** is increased. For example, to undo a tight nut with a spanner, you need a large moment. This can be achieved by applying a large force with a spanner. If a longer spanner is used, a smaller force is needed and it is much easier to undo the nut. The force is applied at a greater distance.

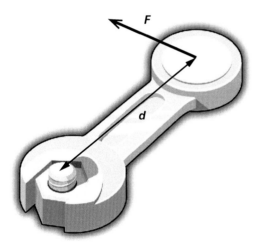

- Examples where moments are important include:
  - door handles
  - see-saws
  - cranes
  - levers.

Note

Doors are difficult to open or close if you pull or push near the pivot (the hinges); the further from the pivot you try to open or close the door, the easier it is. **The greater the distance from the pivot, the smaller the force required.**

Principle of moments

- The **principle of moments** states that, for a beam to be balanced about a pivot, the sum of the clockwise moments equals the sum of the anti-clockwise moments.

> sum of clockwise moments = sum of anti-clockwise moments
> $$F_2 \times d_2 = F_1 \times d_1$$

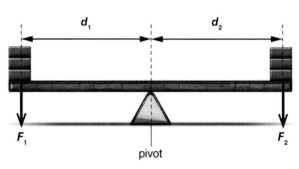

F_1 causes an anti-clockwise turning effect.
F_2 causes a clockwise turning effect

- The following experiment demonstrates the principle of moments.

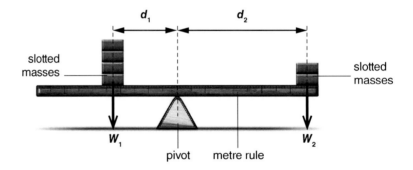

A typical experimental arrangement is shown above.

Apparatus:

- one triangular prism to act as the pivot
- one metre rule
- slotted masses, 100 g each (of weight 1.0 N each).

Method

1. Balance the metre rule on the pivot at the 50.0 cm mark of the rule.

2. Add different masses (of weights) W_1 and W_2 at different distances d_1 and d_2 from the pivot. Carefully adjust the distances d_1 and d_2 until the metre rule balances horizontally.

3. Record the values of W_1, W_2, d_1 and d_2.

4. Repeat steps 2 and 3 several times, with different values for W_1, W_2, d_1 and d_2.

5. For each set of results, calculate the turning moments $W_1 \times d_1$ and $W_2 \times d_2$. You will see that $W_1 \times d_1 = W_2 \times d_2$. The metre rule is balanced when the clockwise moment equals the anti-clockwise moment, thus demonstrating the principle of moments. There is no net turning effect when a body is in equilibrium.
   Remember: Weight is a force, but we use W instead of F in this case.

   The values below are a sample set of results showing that the anti-clockwise moment = the clockwise moment.

W_1/N	d_1/m	W_2/N	d_2/m	$W_1 \times d_1$/Nm	$W_2 \times d_2$/Nm
1.0	0.50	2.0	0.25	0.50	0.50
2.0	0.30	3.0	0.20	0.60	0.60

Note

If there is no net moment and no net force, a system is said to be **in equlibrium**.

Example 1c (viii)

The diagram below shows an experimental set-up for investigating the moment (turning effect) of a force. The metre rule shown is balanced and therefore in equilibrium.

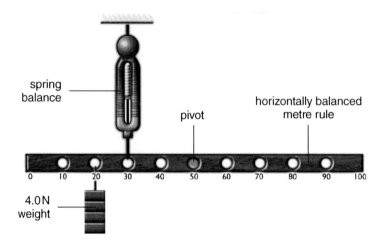

(i) State the equation linking the moment to the force and the perpendicular distance from the pivot.

(ii) Show that the reading on the spring balance is 6.0 N.

(iii) The weight of the rule is 1.7 N. Calculate the force exerted by the pivot on the metre rule.

Answer

(i) moment = force × perpendicular distance from the pivot
or $M = F \times d$

(ii) **Step 1** List all the information in symbol form and change into appropriate and consistent SI units if required.

$F_1 = 4.0\,\text{N}$
$d_1 = 30\,\text{cm} = 0.30\,\text{m}$
$d_2 = 20\,\text{cm} = 0.20\,\text{m}$
$F_2 = ?$

Step 2 The metre rule is balanced, therefore we know that the total clockwise moment must equal the total anti-clockwise moment

$F_2 \times d_2 = F_1 \times d_1$

Step 3 Calculate the answer by putting the numbers into the equation.

$$F_2 \times d_2 = F_1 \times d_1$$
$$F_2 \times 0.20 = 4.0 \times 0.30$$
$$F_2 = \frac{4.0 \times 0.30}{0.20}$$
$$F_2 = 6.0\,N$$

(iii) As the ruler is in equilibrium, the resultant force is zero, and the forces balance, so upward force = downward force

The upward force is 6.0N (reading on the spring balance).

The downward force is the weight + weight of rule:
4.0N + 1.7N = 5.7N

Therefore a 0.30N downward force is required to balance the other forces, and so this is the force exerted by the pivot on the metre rule.

ALWAYS REMEMBER TO STATE THE UNIT FOR CALCULATED QUANTITIES.

○ When there is no resultant force and no resultant moment (turning effect), a system will be **in equilibrium** (balanced).

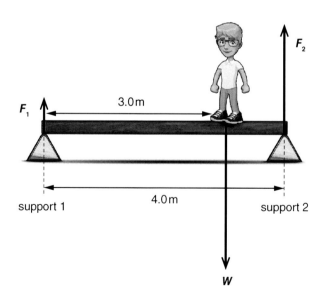

○ In the diagram the beam is in equilibrium
  the upward force = the downward force

and the sum of clockwise = the sum of anti-clockwise
      moments   moments

- If the boy weighs 500N and he stands in the middle of the beam there will be an upward force of 250N at each support (ignoring the weight of the beam itself).

- If the boy stands at the 3.0m mark as shown in the diagram, the upward force at support 2 is greater than that at support 1. This can be shown by the following calculation (which ignores the weight of the beam).

    upward force = downward force

    (1) $\quad F_1 + F_2 = 500\,\text{N}$
    $\quad\quad F_1 = 500 - F_2$

    sum of clockwise moments = sum of anti-clockwise moments

    (2) $\quad F_1 \times d_1 = F_2 \times d_2$
    $\quad\quad F_1 \times 3.0 = F_2 \times 1.0$

    Substituting (1) into (2)
    $(500 - F_2) \times 3.0 = F_2$
    $1500 - 3 \times F_2 = F_2$
    $1500 = 4 \times F_2$
    $F_2 = 375 = 380\,\text{N}$ (to 2 sig. figs)

    ALWAYS REMEMBER TO STATE THE UNIT FOR CALCULATED QUANTITIES.

    The upward force F_2 is 375N, and so the upward force F_1 is 125N.

- The closer the boy stands to support 2, the greater the upward force F_2. If he stands directly above support 2, all of his weight acts down through support 2 and the upward force F_2 is 500N. It follows that F_1 is 0N.

- As the boy moves along the beam from left to right, F_1 decreases from 500N to 0N, and F_2 increases from 0N to 500N.

Section 2 Electricity Units

Section 2a Units

❏ In this section you will come across the following units:
- the **ampere** (A) is the unit of current
- the **coulomb** (C) is the unit of charge
- the **joule** (J) is the unit of energy
- the **ohm** (Ω) is the unit of resistance
- the **second** (s) is the unit of time
- the **volt** (V) is the unit of voltage
- the **watt** (W) is the unit of power.

Section 2b Mains electricity

❏ When electrons flow through a conductor we say there is an electric **current**.

❏ An electric circuit consists of a power supply (which gives energy to the electrons and causes them to move), various components and connectors. The connectors are good conductors, i.e. the electrons can move easily through them from one component to the next.
N.B. There is current in a conductor only if there is a closed circuit with a power supply.

❏ In the home we use electricity for many applications. The moving electrons take energy from the power supply to the circuit components, where it is transferred into a useful form of energy, e.g. a lamp transfers energy to the surroundings by radiation (light).

❏ Examples of electrical devices transferring energy include:
- an electric motor – energy transfers electrically to kinetic energy (which is useful); energy is also transferred by heating and by radiation (sound) to the surroundings (which is not useful)
- a television – the power supply transfers energy electrically to the TV, which transfers energy by radiation (light and sound) to the surroundings and also by heating to the surroundings

- a smartphone – the battery transfers energy electrically to the smartphone, which in turn transfers energy by radiation (light and sound) to the user and also by heating to the surroundings
- an immersion heater – this transfers energy by heating to the water (which is useful) and also to the surroundings (which is not useful).

- [ ] In all energy transfers, some of the energy is always dissipated to the surroundings, e.g. as well as giving out energy by radiation (light), a lamp transfers energy to the surroundings by heating up and this energy is 'wasted'. No component is 100% efficient.

- [ ] All components in an electrical circuit **oppose** current in them, i.e. they have **resistance**.

- [ ] **Resistance** can be thought of as '**electrical friction**'. Current in a wire or component produces a heating effect. This is used to good effect in heaters, toasters, kettles and filament lamps. It happens because the **moving charges** (**electrons**) in the conductor collide with the **ions** inside the conductor, causing resistance. The electrons transfer energy to the ions, which vibrate faster. This heats the wire. (See page 278 for the definition of an ion.)

- [ ] **Direct current** (**d.c.**) is the flow of charge in one direction.

- [ ] **Alternating current** (**a.c.**) is the flow of charge backwards and forwards. It changes direction many times every second.

- [ ] The **mains** supply (electricity from wall sockets) is always **a.c.** A **battery** supply is always **d.c.**

Power and energy

- [ ] Energy transfer has the symbol E, and is the energy of the moving charges in an electrical transfer. Its unit is the **joule** (J).

- [ ] All electrical equipment has a **power rating**. The power rating indicates how much energy is supplied to the device each second.

2b Mains electricity

❑ Power has the symbol **P**, and is defined as the rate of transfer of energy. Its unit is the **watt** (W), e.g. 2000 J of energy is transferred to the water by heating per second by a 2000 W electric kettle.

1 W = 1 J/s

$$\text{power} = \frac{\text{energy transferred}}{\text{time taken}}$$

$$P = \frac{E}{t}$$

P = power (W)
E = energy (J)
t = time taken (s)

❑ The power of an electrical component depends on the current through the component and the voltage applied to it. The voltage of an electricity supply is an indication of the energy it provides, measured in volts (V). The voltage of the mains supply in many countries is 230 V.

❑ The power of an electrical component can be calculated by multiplying the current value **I** by the voltage value **V**.

power = current × voltage

$$P = I \times V$$

P = power (W)
I = current (A)
V = voltage (V)

❑ As power is defined as the rate of transfer of energy:

$$P = \frac{E}{t} \quad \Rightarrow \quad E = P \times t$$

and as $P = I \times V$, so we obtain the relationship

energy transferred = current × voltage × time

$$E = I \times V \times t$$

E = energy transferred (J)
I = current (A)
V = voltage (V)
t = time (s)

Example 2b (i)

(i) State the equation linking power, energy transferred and time.
(ii) Calculate how much energy is transferred by a 100W lamp in 2.0 minutes.

Answer

(i) power = $\dfrac{\text{energy transferred}}{\text{time taken}}$ or $P = \dfrac{E}{t}$

(ii) **Step 1** List all the information in symbol form and change into appropriate and consistent SI units if required.

P = 100 W
t = 2.0 minutes = 2 × 60 = 120 s
E = ?

Step 2 Rearrange the equation.

$P = \dfrac{E}{t}$ ⇨ $E = P \times t$

Step 3 Calculate the answer by putting the numbers into the equation.

E = 100 × 120 = 12 000 J

Example 2b (ii)

The current in an electric cooker that is connected to the mains supply at 230 V is 8.0 A.

(i) State the equation linking power, current and voltage.
(ii) Calculate the power of the cooker.
(iii) Calculate the energy transferred if the cooker is on for 30 minutes.

Answer

(i) power = current × voltage or $P = I \times V$

(ii) **Step 1** List all the information in symbol form and change into appropriate and consistent SI units if required.

V = 230 V
I = 8.0 A
P = ?

Step 2 Calculate the answer by putting the numbers into the equation.

P = 8.0 × 230 = 1840 W

(iii) **Step 1** List all the information in symbol form and change into appropriate and consistent SI units if required.

P = 1840 W
t = 30 minutes = 30 × 60 s = 1800 s
E = ?

Step 2 Calculate the answer by putting the numbers into the equation.

E = 1840 × 1800 = 3 312 000 = 3 300 000 J
(to 2 sig. figs)

ALWAYS REMEMBER TO STATE THE UNIT FOR CALCULATED QUANTITIES.

Example 2b (iii)

(i) State the equation linking energy transferred to voltage, current and time.
(ii) Calculate the energy transferred by a 230 V hairdryer, running on a current of 7.5 A, that is left on for 8.0 minutes.

Answer

(i) energy transferred = current × voltage × time or $E = I \times V \times t$

(ii) **Step 1** List all the information in symbol form and change into appropriate and consistent SI units if required.

V = 230 V
I = 7.5 A
t = 8.0 minutes = 8 × 60 = 480 s
E = ?

Step 2 Calculate the answer by putting the numbers into the equation.

E = 7.5 × 230 × 480 = 828 000
= 830 000 J (to 2 sig. figs)

ALWAYS REMEMBER TO STATE THE UNIT FOR CALCULATED QUANTITIES.

2 Electricity

Potential hazards

- You will get an electric shock if you touch a live wire. The live wire carries the current at high voltage to the appliance. If you touch it, the current will pass through you. To prevent this, cables are insulated by covering or separating them with a material that does not conduct electricity. Such a material, for example rubber or plastic, is called an **insulator**.

- **Frayed cables** can lead to the insulation around the wires becoming damaged, which might lead to the live wire becoming exposed and dangerous. Accidentally touching this would cause an electric shock.

- In addition, exposed cables might touch each other, leading to a short circuit – a low-resistance connection between two conductors that causes a large current. **Overheated cables** could lead to a fire; this is particularly hazardous due to the fact that much of the wiring in any building is hidden in the walls and underneath floors.

- Using mains electrical appliances in **damp** conditions could lead to a person becoming directly connected to the live mains supply (because impure water conducts), which would cause serious shock and possible death.

Safety measures

- **Fuses** and **circuit breakers** are designed to break the circuit if a fault occurs that causes too much current to flow.

- A **fuse** is a deliberate weak link in a circuit for safety. It has the symbol:

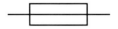

- Fuses are found in plugs and there may be a main fuse in the meter cupboard where the electricity supply enters the building.

- If there is **too much current**, the fuse wire **melts** and **disconnects the circuit**; it protects the flex (the flexible cable between the plug and appliance) and helps prevent electrical fires.

- The **fuse rating** is always **slightly higher** than the current in the appliance. For example, a 10A appliance should use a 13A fuse. If it used a fuse with a rating less than 10A, the appliance would not even switch on. The fuse would melt immediately on switching on the appliance.

- Older buildings also have fuses protecting each circuit. Modern buildings are protected by circuit breakers instead.

- A **circuit breaker** is an automatic safety switch. It springs open (**trips**) if there is too much current. This switch can easily be reset once the fault is corrected.

> **Top Tip**
>
>
>
> Electrical metal wires get hot when there is a current in them. **Thicker wires** can be used to **reduce** this heating effect, as the greater cross-sectional area reduces their resistance. But using **thicker insulation** only disguises the heating effect; the wires inside could still overheat.

Other safety measures

N.B. Colours and plug designs may vary in different countries.

- In the UK, electrical appliances are connected to the mains supply using a three-pin plug.

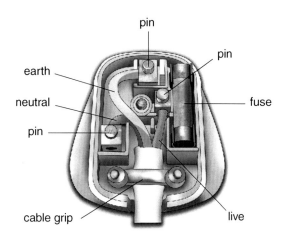

- The **live** wire in a flex is normally brown and is connected to the right-hand pin of a plug when looking down on the plug as shown in the previous diagram. It is linked to the fuse.
  The **neutral** wire is normally blue and is connected to the left-hand pin.
  The green and yellow **earth** wire is connected to the top pin of the plug.
  Colour coding allows us to safely identify wires.

- Many electrical appliances, e.g. washing machines, cookers, and refrigerators, have metal casings. If the live wire were to become loose and make contact with the **external metal casing** of an appliance, the appliance would become dangerous. An **earth wire** provides the current with a **low resistance path** from the metal casing of the device to earth (the ground). A large current is produced and the **fuse blows**. The circuit is **disconnected**, preventing a shock.

- Some appliances, such as vacuum cleaners and electric drills, do not need an earth wire as they have a non-conducting (or insulating) casing such as **plastic**. This casing can never become live and give an electric shock. Appliances like this are said to be **double insulated**. Such appliances are marked with this symbol:

- Some UK electrical appliances are fitted with a two-pin plug. Two-pin plugs can only be used safely with appliances that are double insulated.

Example 2b (iv)
(i) State the equation linking electrical power, current and voltage.
(ii) Determine the appropriate fuse rating for the following appliances. You can choose from a 3A, 5A or 13A fuse.
   (a) A 2.5kW electric kettle operating on a 230V supply.
   (b) An 800W electric drill operating on a 230V supply.
   (c) A 40W computer operating on a 16V supply.

Answer

(i) power = current × voltage or $P = I \times V$

(ii) **Step 1** List all the information in symbol form and change into appropriate and consistent SI units if required.

    (a) $P = 2.5\,\text{kW} = 2500\,\text{W}$
        $V = 230\,\text{V}$
        $I = ?$

    (b) $P = 800\,\text{W}$
        $V = 230\,\text{V}$
        $I = ?$

    (c) $P = 40\,\text{W}$
        $V = 16\,\text{V}$
        $I = ?$

Step 2 Rearrange the equation:

$$P = I \times V \quad \Rightarrow \quad I = \frac{P}{V}$$

Step 3 Calculate the answer by putting the numbers into the equation.

    (a) $I = \dfrac{2500}{230} = 10.9\,\text{A}$

You would need a 13 A fuse.

    (b) $I = \dfrac{800}{230} = 3.48\,\text{A}$

You would need a 5 A fuse.

    (c) $I = \dfrac{40}{16} = 2.5\,\text{A}$

You would need a 3 A fuse.

ALWAYS REMEMBER TO STATE THE UNIT FOR CALCULATED QUANTITIES.

Section 2c   Energy and voltage in circuits

Charge and current in a circuit

- Electric current can pass in a **conductor** because charges are free to move. Current in metals is due to a flow of negatively charged electrons (see page 63).

- **Current** has the symbol I, and its unit is the **ampere** (A).

- **Charge** has the symbol Q, and its unit is the **coulomb** (C).

- **Electric current** is the **rate of flow** of electric charge.

> **Note**
>
>
> The charge on an electron is 1.6×10^{-19} C, which means 1.0 C of charge has 6.25×10^{18} **electrons**, a huge number.

- Current, charge and time are related by the following equation:

    charge = current × time

    $$Q = I \times t$$

    Q = charge (C)
    I = current (A)
    t = time (s)

    By rearrangement:

    $$\text{current} = \frac{\text{charge}}{\text{time}} \quad \text{or} \quad I = \frac{Q}{t}$$

Example 2c (i)

(i) State the equation linking current, charge and time.
(ii) Calculate the current in a wire when 720 C of charge is transferred in 4.0 minutes.

Answer

(i) charge = current × time or $Q = I \times t$

(ii) **Step 1** List all the information in symbol form and change into appropriate and consistent SI units if required.

$Q = 720\,C$
$t = 4.0 \text{ minutes} = 4.0 \times 60\,s = 240\,s$
$I = ?$

Step 2 Rearrange the equation.

$Q = I \times t \quad \Rightarrow \quad I = \dfrac{Q}{t}$

Step 3 Calculate the answer by putting the numbers into the equation.

$I = \dfrac{720}{240} = 3.0\,A$ (to 2 sig. figs)

ALWAYS REMEMBER TO STATE THE UNIT FOR CALCULATED QUANTITIES.

❑ Although electrons flow from the negative terminal of a power supply round the circuit to the positive terminal, scientists use the convention that the direction of electric current is from positive to negative.

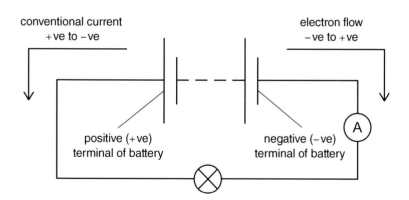

Note

Electrons flow from negative to positive.
Conventional current is from positive to negative.

Remember: When we talk about current we mean conventional current.

2 Electricity

Voltage and energy

- **Voltage** supplies electrons energy to move around a circuit. It has the unit **volt** (V).

- The relationship between energy transferred, charge and voltage is given by the following equation:

  energy transferred = charge × voltage

  $$E = Q \times V$$

  E = energy transferred (J)
  Q = charge (C)
  V = voltage (V)

- **One volt** is defined as one joule per coulomb. **1 V = 1 J/C**

- The battery or power supply gives energy to the charges, which carry the energy electrically round a circuit to the various components. The stated voltage of the battery or power supply is the energy it can provide per coulomb of charge.

- As charge flows through a circuit component such as a **resistor**, it transfers energy, i.e. the charge has more energy as it starts to flow through the component than when it leaves. The electrical energy is transferred by heating, by radiation (light and sound), etc. Hence there is a voltage drop across a component.

- All components in an electrical circuit have **resistance** (see pages 74-80). Generally wires with a very low resistance, e.g. copper wire, are used to connect components so the charges flow easily between them. Materials with a higher resistance, e.g. nichrome or carbon, are used to make resistors.

- Resistance has the symbol R, and its unit is the **ohm** (Ω).

- When the total resistance in a circuit increases and the voltage of the source remains constant, the current in the circuit decreases.

Example 2c (ii)

(i) State the equation linking energy transferred, voltage and charge.

(ii) Calculate how much energy a 12 V battery will give to 5.0 C of charge.

Answer

(i) energy transferred = charge × voltage   or $E = Q \times V$

(ii) **Step 1** List all the information in symbol form and change into appropriate and consistent SI units if required.

$Q = 5.0\,C$
$V = 12\,V$
$E = ?$

Step 2 Calculate the answer by putting the numbers into the equation.

$E = 5.0 \times 12 = 60\,J$

ALWAYS REMEMBER TO STATE THE UNIT FOR CALCULATED QUANTITIES.

Note

Voltage is defined as the energy transferred by each coulomb of charge.

$$V = \frac{E}{Q}$$

1 volt = 1 joule/coulomb or **1 V = 1 J/C**

2 Electricity

Electric circuit symbols

It is important to become familiar with the following electric circuit symbols for the rest of this topic. You have to be able to recognise and draw them.

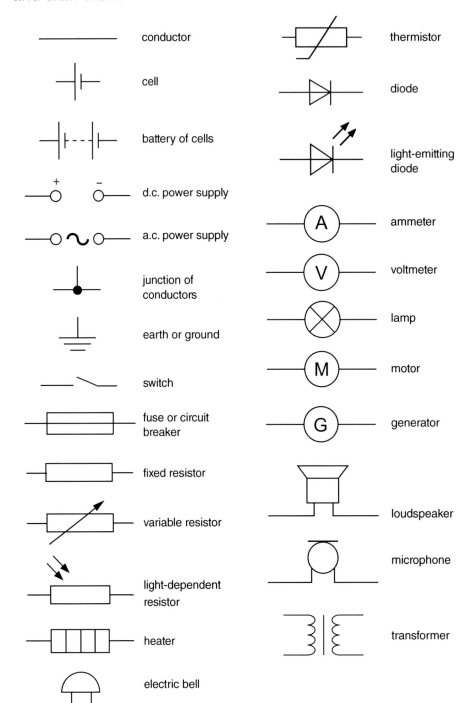

Series and parallel circuits

❑ Components in an electrical circuit can be connected one after another (in **series**) or side by side (in **parallel**).

Series circuits

❑ There is **only one path** for the current in a **series circuit**.

❑ The **current is the same at all points** in a series circuit:

$$I = I_1 = I_2$$ I = supply current as measured by an ammeter

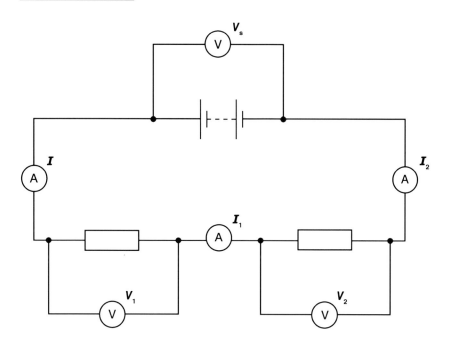

(The various meters are assumed not to be part of the circuit as they are simply there to measure values and have no effect on the circuit.)

❑ The **supply voltage** V_s is equal to the **sum of the voltages** across each **individual component** in a series circuit:

$$V_s = V_1 + V_2$$ V_s = supply voltage

❑ The combined **total resistance** of resistors connected in a series circuit is the sum of the individual resistances. The current in the series circuit depends on the combined resistance of all of the components.

❑ A practical example of a series circuit is a simple flashlight consisting of a battery, a switch and a lamp.

Parallel circuits

- In a **parallel circuit** there is **more than one path** for the current. In the following diagram, **R_1** and **R_2** are resistors.

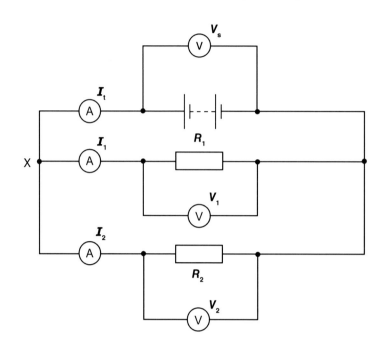

(The various meters are assumed not to be part of the circuit as they are simply there to measure values and have no effect on the circuit.)

- Current is **conserved at a junction**. This means that the total current entering a junction (such as X) is the same as the total current leaving the junction. If this didn't happen you would either get a build-up of charge at a junction or you would have to create charge from nowhere, and neither can happen.

- The **sum of the currents** in the individual parallel branches is equal to the **current drawn from the supply I_t**:

$$I_t = I_1 + I_2$$

I_t = total supply current as measured by the ammeter in series with the battery

- The current in a parallel circuit is larger in the branch containing the battery than in the other branches.

- If R_1 is greater than R_2, then I_1 will be less than I_2.

- The **voltages across individual components** in a parallel circuit are **equal**:

$$V_s = V_1 = V_2$$

V_s = supply voltage
V_1 and V_2 = voltage across resistors

Ammeters and voltmeters

- An **ammeter** has **very low resistance** to current and so does not affect the current in the circuit: it simply measures it.

- An ammeter is always placed in **series** with other components in a circuit.

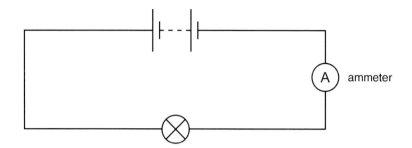

- An analogue ammeter gives a reading with a needle moving across a scale; a digital ammeter has a numerical reading displayed.

- The analogue ammeter scale below shows a current of 0.50 A.

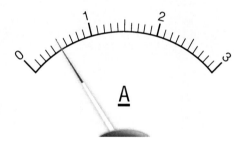

- The digital ammeter display below shows a current of 1.26 A.

- [ ] A **voltmeter** can similarly be either analogue or digital.
- [ ] The analogue voltmeter below shows a voltage reading of 2.40 V.

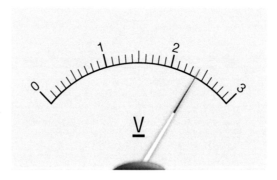

- [ ] The digital voltmeter below shows a voltage reading of 3.32 V.

- [ ] A voltmeter is always connected in **parallel** with the component whose voltage is being measured, as shown in the diagram. They do not affect the circuit. Voltmeters have **very high resistance**.

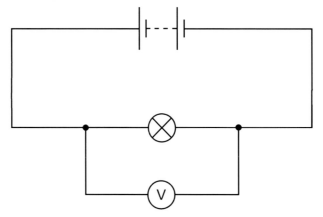

voltmeter measuring the voltage across the lamp

2c Energy and voltage in circuits

Understanding electric currents

❑ It can be difficult to visualise what is happening in an electric circuit because electric current is invisible. But we can compare an electric circuit with a water circuit.

Imagine a circuit consisting of water pipes and a pump moving the water round.

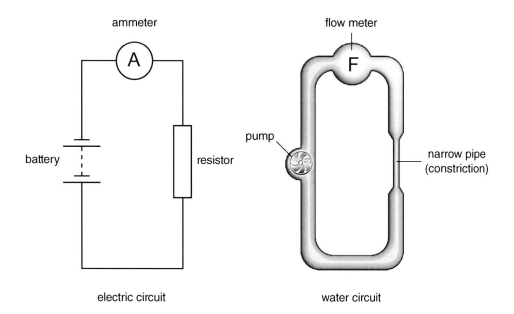

electric circuit   water circuit

- The pump gives the water energy just as a battery gives the electrons energy.
- A flow meter would measure the rate of flow of the water just as an ammeter measures the rate of flow of charge.
- A constriction (narrowing) in the pipe reduces the flow rate (slows the water down) just as a **resistor** reduces the current (slows the charges).

N.B. The flow rate is reduced both before and after the constriction, i.e. the narrowing affects the flow rate throughout the whole circuit.
Similarly, a resistor affects the current through the whole of a series circuit.

Note

 Resistance is a measurement of a conductor's **opposition** to the **flow of electric current**. It is measured in ohms.

Series circuit with identical lamps

- In a **series** circuit as shown below, the lamps will light if the switch is closed. Lamps can be used to indicate if there is current in a circuit.

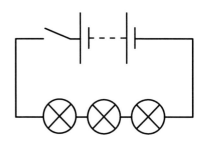

- In a series circuit, if one lamp blows the others will **go out** (e.g. some festival lights).

- In a series circuit, the lamps **cannot** be switched on and off **independently**. When the switch is open there can be no current in the circuit.

- In a series circuit, all lamps **share** the supply voltage from the battery, so identical lamps are **equally bright** when the switch is closed.
  Remember: Current is the **same** at **all points** in a series circuit.

- If more lamps are added in a series circuit, there is more resistance and so less current and less voltage per lamp. The more lamps there are, the dimmer they **all** become.

Parallel circuit with identical lamps

- In a **parallel** circuit as shown below, if one lamp blows the others will remain **on** provided all the switches are closed.

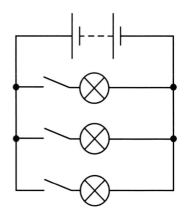

2c   Energy and voltage in circuits

- ❏ In a parallel circuit, the lamps **can** be switched on and off independently. If one switch is open it will affect only the lamp in the same branch of the circuit.

- ❏ In a parallel circuit, all lamps have the **same voltage** as the **battery**.

- ❏ All the lamps in a parallel circuit are **equally bright** when all of the switches are closed, because they have the same voltage and they are identical, so have equal resistance and equal current.

Note

The **brightness** of the lamps **depends** on **power** ($P = I \times V$). Lamps in series in a circuit are dimmer than they would be if they were in parallel with the same supply, because there is less voltage per lamp and less current in the lamp because of the higher resistance of the circuit. This in turn causes the power used per lamp to be less.

- ❏ The **total current**, found by adding the current through each lamp, **adds** up to the **supply** current.

- ❏ If more lamps are added in parallel, the brightness of all lamps stay the same, just as bright as before adding the extra lamps.

- ❏ A practical example of a parallel circuit is the domestic wiring in your home. You can switch one light off without switching others off, and you can switch the television off without switching off the computer.

Top Tip

The advantages of parallel circuits are:
- When one component such as a lamp goes off, the others stay on (because there is more than one path for the current).
- Each component such as a lamp has the same voltage as the supply voltage.
- Each component can be switched on and off independently.

Ohm's Law

- **Voltage**, **current** and **resistance** are related by the equation:

  voltage = current × resistance

  $$V = I \times R$$

  V = voltage (V)
  I = current (A)
  R = resistance (Ω)

  By rearrangement:

  $$\text{resistance} = \frac{\text{voltage}}{\text{current}} \quad \text{or} \quad R = \frac{V}{I}$$

- This equation **defines the resistance** of a component as the ratio V/I and it is constant for metallic conductors provided the temperature is constant.

- So if the **voltage** across a resistor **doubles**, the **current** in it **doubles**.

- If the resistance of a conductor **remains constant**, a graph of voltage against current is a straight line through the origin. The voltage is proportional to the current ($V \propto I$). The **gradient** of the voltage–current graph in this case is the value of the resistance of the conductor.

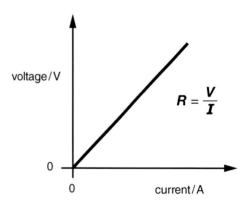

Note

For a given metallic conductor, **V/I is constant provided that its temperature is constant**. This is known as Ohm's Law. The conductor is sometimes described as ohmic.

❏ If a device has a fixed resistance R, then you can calculate V or I if you know the other value from $V = I \times R$.

Example 2c (iii)

(i) State the equation linking current, voltage and resistance.
(ii) Calculate the current in a resistor that has a 2.0 kΩ resistance and is connected across a 230 V supply voltage.

Answer

(i) voltage = current × resistance   or $V = I \times R$

(ii) **Step 1**   List all the information in symbol form and change into appropriate and consistent SI units if required.

$V = 230\,\text{V}$
$R = 2.0\,\text{k}\Omega = 2000\,\Omega$
$I = ?$

Step 2   Use and rearrange the correct equation.

$$V = I \times R \quad \Rightarrow \quad I = \frac{V}{R}$$

Step 3   Calculate the answer by putting the numbers into the equation.

$$I = \frac{V}{R} = \frac{230}{2000} = 0.12\,\text{A} \quad \text{(to 2 sig. figs)}$$

ALWAYS REMEMBER TO STATE THE UNIT FOR CALCULATED QUANTITIES.

Example 2c (iv)

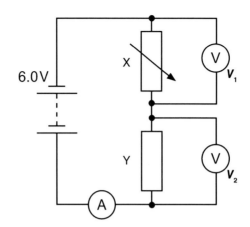

The circuit above shows two components X and Y connected in series to a 6.0 V battery. An ammeter measures the current through the components and a voltmeter is connected across each component.
The reading on the ammeter is 1.2 A and the reading V_1 is 3.6 V.

(i) What are components X and Y?
(ii) What is the equation linking current, voltage and resistance?
(iii) Calculate the resistance of X.
(iv) Give the reading V_2 and explain your answer.
(v) Calculate the resistance of Y.

Answer

(i) X is a variable resistor and Y is a fixed resistor.

(ii) voltage = current × resistance   or $V = I \times R$

(iii) **Step 1** List all the information in symbol form and change into appropriate and consistent SI units if required.

$I = 1.2\,A$
$V = 3.6\,V$
$R = ?$

Step 2 Rearrange the equation.

$V = I \times R \quad \Rightarrow \quad R = \dfrac{V}{I}$

2c Energy and voltage in circuits

Step 3 Calculate the answer by putting the numbers into the equation.

$$R = \frac{3.6}{1.2} = 3.0\,\Omega \quad \text{(to 2 sig. figs)}$$

(iv) $V_s = V_1 + V_2$

The total voltage supplied by the battery is 6.0 V and V_1 reads 3.6 V so the reading on V_2 is 6.0 − 3.6 = 2.4 V.

(v) **Step 1** List all the information in symbol form and change into appropriate and consistent SI units if required.

$I = 1.2\,\text{A}$
$V = 2.4\,\text{V}$
$R = ?$

Step 2 Rearrange the equation.

$$V = I \times R \quad \Rightarrow \quad R = \frac{V}{I}$$

Step 3 Calculate the answer by putting the numbers into the equation.

$$R = \frac{2.4}{1.2} = 2.0\,\Omega \quad \text{(to 2 sig. figs)}$$

ALWAYS REMEMBER TO STATE THE UNIT FOR CALCULATED QUANTITIES.

Determining the resistance of an unknown resistor

❏ The following experiment describes how to determine the resistance of an unknown resistor.

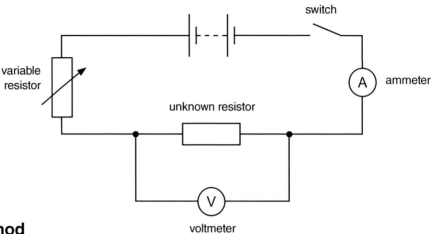

Method

1. Set up the circuit as shown.

2. Vary the current by varying the resistance of the variable resistor.

3. Record the current I through the unknown resistor and the voltage V across it.

4. Repeat for several different values of I and V and construct a table of results.

5. Plot a graph of V against I.

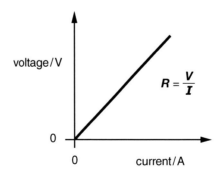

N.B. If any of the results are anomalous (unexpected and not fitting on the line), repeat them to check.

Calculations to be made

Resistance of the unknown resistor.

$$V = I \times R \implies R = \frac{V}{I}$$

which is equal to the gradient of the V–I graph.

❏ We expect the graph to be a straight line passing through the origin, because Ohm's Law tells us $V \propto I$ if the resistance remains constant. If the wire heats up during the course of the experiment, its resistance will increase and the graph will be a curve with an increasing gradient.

Improvements to the experiment
The experiment could be improved by:
1. **switching off** the circuit **between readings** to reduce the heating effect
2. carrying out the experiment at a lower supply voltage to reduce the heating effect.

Assumptions in the experiment
We assume that the changing temperature has no effect on the resistor's resistance. A straight-line graph through the origin will confirm this.

Resistance of a filament lamp

❏ A similar experiment can be carried out replacing the unknown resistor by a filament light bulb.

❏ Look at the graph of voltage against current for a filament lamp shown below. You can tell the resistance is changing because the graph is now a curve, so the gradient is changing.
The resistance increases as the current increases because the filament becomes hotter. This means that the voltage does not increase uniformly as it would in an ohmic resistor. You can calculate the resistance at any point on the curve by dividing the voltage value at that point by the corresponding current value.

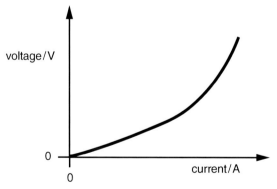

R increases as voltage increases

Resistance of a diode

❑ A **diode** is a component that only allows current to pass through it in one direction. A current–voltage graph for a diode will look like this:

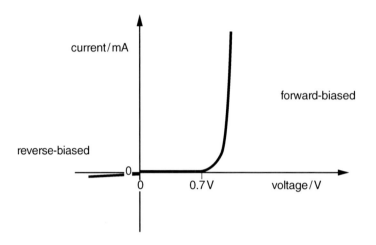

N.B. This is a current–voltage graph, not a voltage–current graph.

❑ The graph shows that there is effectively zero current when a voltage is applied in the reverse-biased direction, and in the forward-biased direction a voltage of 0.7 V is required before there is current in the circuit. Then as the voltage increases, the current increases at an increasing rate.

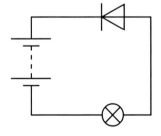

No current
The diode is reverse-biased.
The diode has an infinitely high resistance.

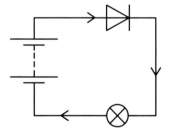

There is a current
The diode is forward-biased.
The diode has a low resistance.

❑ The process of changing a.c. into d.c. is known as **rectification**.

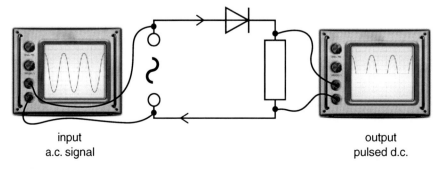

input
a.c. signal

output
pulsed d.c.

2c Energy and voltage in circuits

❏ Diodes are used to change a.c. into d.c., for example in battery chargers where the charging current must be in the correct direction at all times.

❏ Diodes are found in computers and television sets because microchips require d.c.

Current–voltage graphs

❏ In summary, the relationship between current and voltage in an ohmic conductor, a filament lamp and a diode can be shown by these current–voltage graphs, including reverse voltage.

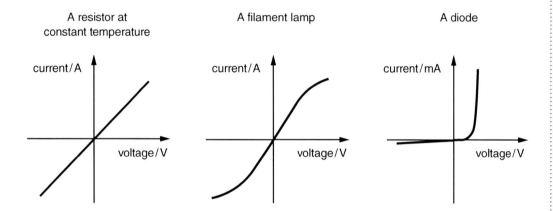

Other circuit components

❏ Many real-life circuits are used for controlling devices.

❏ All **control systems** have an input, a processor and an output.

❏ An input sensor sends a signal to a processor, which controls the flow of power to an output device.

2 Electricity

- Devices that **transfer** energy from one form to another are called **transducers**. The components below are all transducers.

Input devices	Output devices
light-dependent resistor (LDR)	light-emitting diode (LED)
microphone	lamp
thermistor	buzzer
variable resistor	loudspeaker
pressure switch (switch operated by pressing it)	electric motor

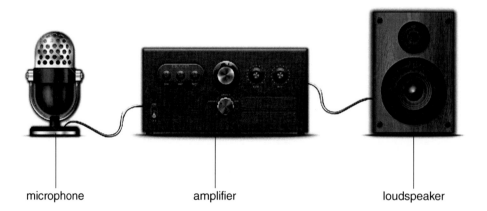

microphone   amplifier   loudspeaker

- Energy from a source such as a musical instrument is received by the microphone as sound (a form of radiation). This energy is transferred electrically to the amplifier, which amplifies the signal by adding some of its own energy before transferring it to the loudspeaker, where it is transferred to the surroundings by radiation (sound).

- A **variable resistor** controls the current and often the voltage of circuit components. Its symbol is:

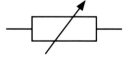

- A **thermistor** is a resistor with a resistance that depends on temperature. Its symbol is:

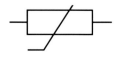

- As the **temperature increases**, the **resistance of a thermistor decreases** and vice versa.

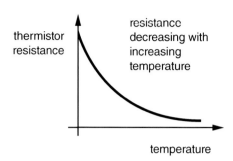

- The resistance of a **light-dependent resistor** (**LDR**) varies according to the amount of light falling on it. Its symbol is:

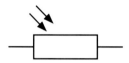

- As the **light intensity increases**, the **resistance of an LDR decreases** and vice versa.

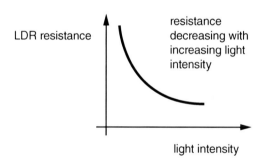

- A **diode** allows current to pass in **one direction** only. The current can only flow in the direction of the 'arrowhead'. Its symbol is:

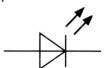

- A **light-emitting diode** is a diode that glows when current passes through it. Therefore an LED can be used to indicate that current is present in a circuit. The symbol for a light-emitting diode is:

Section 2d Electric charge

- All matter contains particles called **electrons**, which have an electric charge. When these electrons are transferred from the surface of one material to the surface of another, and stay there, they produce an **electrostatic charge** ('static' means not moving).

- Examples of electrostatic charges include:
    - rubbing a balloon on your hair and then sticking it to a wall
    - rubbing a comb on your jumper and then bringing it near to small pieces of paper; the paper jumps and sticks to the comb
    - rubbing a polythene rod and bringing it close to a slow trickle of water; the water is attracted to the polythene so the trickle bends.

- Some materials are easier than others to charge by rubbing.

- **Electrical insulators** such as rubber, plastic and glass (non-metals) charge easily.

- When electrons are added to or removed from an insulator's surface by rubbing, the charges stay on the surface because they are **not free to move** through the material.

- Silver, copper and gold are examples of **electrical conductors** (they are metals). Charge passes through them easily.

- Metals are good conductors because they contain **electrons that are free to move**. They are **difficult to charge** by rubbing because the electrons keep moving and the charge flows to earth.
  N.B. The term 'earth' means exactly what it says – the Earth. Earthing something or 'grounding' it means to make a direct electrical connection to the Earth.

Examples of conductors and insulators

Conductors	Insulators
silver	rubber
gold	glass
copper	diamond
aluminium	dry paper
steel	oil
iron	plastic
sea water	pure water
lemon juice	cotton
human body	wool

○ All atoms are made up of three kinds of particles, called **electrons**, **protons** and **neutrons**.

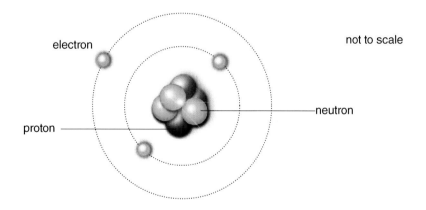

○ **Protons** and **neutrons** are found in the densest part of the atom known as the **nucleus**, whereas **electrons** are found **orbiting** the nucleus.

○ **Protons** are **positively** charged, **electrons** are **negatively** charged and **neutrons** have **no charge**.

○ Atoms usually have **equal** numbers of electrons and protons, making the **resultant charge** of the atom **zero**.

- When two different insulators are rubbed together, electrons may be transferred from one insulator to the other. This leaves one insulator **positively charged** (having lost electrons) and the other one **negatively charged** (having gained electrons).

- **Protons are not** transferred; it is **only electrons** that are transferred.

Production of electrostatic charges

- Electrostatic charges can be produced in the following ways.

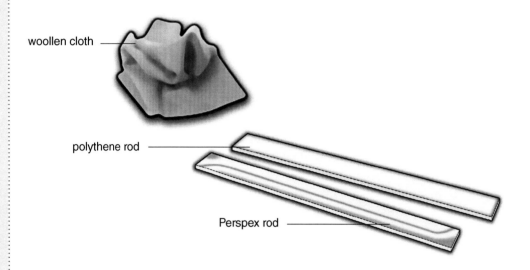

1. Rub the polythene rod with the woollen cloth. This transfers electrons from the wool to the polythene, leaving the polythene **negatively charged** (it gains electrons) and the wool **positively charged** (it loses electrons).

2. Rub the Perspex rod with the woollen cloth. This transfers electrons from the Perspex to the wool, leaving the wool **negatively charged** (it gains electrons) and the Perspex **positively charged** (it loses electrons).

Note

Only **electrons** move when charges transfer by rubbing: the **protons do not**.

If an object **gains electrons** it becomes **negatively charged** and if an object **loses electrons** it becomes **positively charged**.

Forces between charges

- Electric charges exert forces on one another.

- **Like charges** (++ or --) **repel** and **unlike charges** (-+ or +-) **attract**.

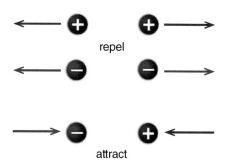

- This can be demonstrated by rubbing two insulating rods with a woollen cloth. Hang one (A) with a piece of cotton as shown, and bring the other (B or C) close.
    - If both rods have the same charge, B will repel A, and A will rotate away from B.
    - If the rods have opposite charges, C will attract A, and A will rotate towards C.

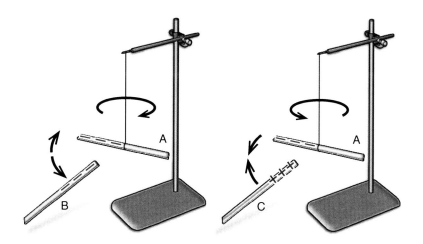

Top Tip

Charged objects will sometimes attract **uncharged objects**. If you put a positively charged balloon close to some small pieces of paper, it attracts them. The pieces of paper could be either neutral or negatively charged; you can't tell whether the paper is charged or not, so attraction is not a true test of whether an object is charged or not.

Top Tip

The **true test** for charge is **repulsion**. If you put a positively charged balloon close to an object such as a rod hanging from a thread, and it repels, the rod is definitely charged and has a positive charge.

Detection of electrostatic charges

- A **gold leaf electroscope** can be used to detect if an object is charged. The metal cap is connected to a metal rod and to a strip of very thin gold leaf, which is hinged so it can move.

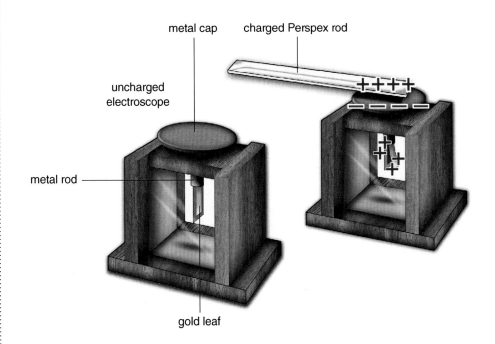

- If you put a positively charged Perspex rod near the cap of the electroscope (**but not touching it**), the gold leaf rises up, away from the metal rod.

- This happens because the electrons in the electroscope's metal rod and gold leaf are attracted towards the Perspex rod. This leaves the cap with a negative charge, and a positive charge on the metal rod and gold leaf (as the free electrons move towards the cap). The positive charges on the metal rod and gold leaf repel each other and so the leaf rises.

Effects of electrostatic charging

- The human body is a conductor of electricity, and if it becomes charged, the charge normally flows to earth. But if you are wearing shoes with rubber soles (remember, rubber is an insulator), the charge is not able to flow to earth and the body can remain charged.

- This can happen when you walk across a carpeted floor in rubber-soled shoes; you will become charged because of the friction between your shoes and the carpet. If you touch a metal radiator or a metal door handle, the charge from your body will jump across to get to earth via the radiator or handle. You will feel a small electric shock.

- If you wear a woollen pullover over a nylon shirt, they will cling together, and as you remove the pullover you might see a spark or get a shock.

- A Van de Graaff generator is a machine that uses a moving insulating belt to accumulate static charge onto a large metal dome. In the diagram a girl stands on an insulating block and touches the dome as it is being charged. She becomes charged. Each strand of her hair will become charged and, because like charges repel, the strands of her hair will repel each other and separate.

2 Electricity

- Sometimes the effects of a build-up of electrostatic charge may be **dangerous**.
    - When adding fuel to a car, the fuel (an insulator) flows through a pipe (also an insulator). Friction between the two creates an electrostatic charge, and when the nozzle is brought close to the fuel tank a spark can jump between the two. This can be avoided by earthing the pipe so that electrons can flow down a wire, instead of charging the pipe.
    - A similar danger is present when an aircraft or a tanker is re-fuelled, and this is a major problem because they are re-fuelled at high speed and the friction between the fuel and the pipe is great. The hoses are earthed to prevent a spark igniting the fuel.

- Electrostatic charges can be useful as well.
    - An electrostatic paint gun is charged, which means the paint is also charged. As the paint leaves the gun, the individual particles of paint repel each other and separate, producing a fine spray, so that the paint will cover a wider area more evenly than if a paintbrush was used.
    - A photocopier uses electrostatic charge to produce a copy of an image. The image, consisting of light and dark areas, is projected onto a positively charged drum. The toner (ink) is negatively charged, and is attracted in varying degrees to the drum. The drum then rotates against a piece of paper and transfers the toner to reproduce the image.
    - Inkjet printers use static electricity to guide a tiny droplet of ink to the correct place on a piece of paper. The droplets are forced through a fine nozzle and become charged. They then pass between charged metal plates and are deflected. Each droplet in turn is deflected to a particular place on the paper.

Section 3 Waves

Section 3a Units

☐ In this section you will come across the following units:
- the **degree** (°) is the unit of angle
- the **hertz** (Hz) is the unit of frequency
- the **metre** (m) is the unit of distance
- the **second** (s) is the unit of time
- the **metre/second** (m/s) is the unit of speed or velocity.

Section 3b Properties of waves

☐ **Waves transfer energy and information** from one place to another. They do **not** transfer matter. Consequently, waves can be used to carry signals from one place to another without transferring matter.

☐ Waves are produced by vibrations.

☐ Waves have repeating patterns.

☐ **Wave terms**:

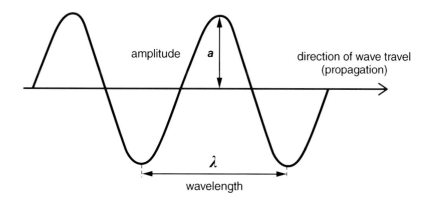

- **amplitude a** – the maximum distance of the wave from the **central equilibrium position**
- **wavelength** λ – the distance from any point on a wave to the next (adjacent) similar point on the wave, e.g. peak to peak or trough to trough.

> *Note*
>
> Waves transfer energy; waves do not transfer matter (particles). The larger the amplitude, the greater the energy the wave transfers.

- Wave **frequency** is defined as the **number of wavelengths** produced or passing a specific point **per second**; it has the symbol *f* and its unit is the **hertz** (Hz).

- The **time period** of a wave is the time taken for one wavelength to pass a specific point; it has the symbol *T* and its unit is the **second** (s).

- Frequency and time period are related by the equation:

$$\text{frequency} = \frac{1}{\text{time period}}$$

$$f = \frac{1}{T}$$

T = time period (s)
f = frequency (Hz)

- The **speed** of a wave *v* is the distance travelled by a wave in one second.

- The **wave equation** is:

wave speed = frequency × wavelength

$$v = f \times \lambda$$

v = speed (m/s)
f = frequency (Hz)
λ = wavelength (m)

Types of waves

- There are two principal types of wave: **longitudinal** waves and **transverse** waves. Most waves are transverse waves, only a few are longitudinal.

Longitudinal waves

- This type of wave can be shown by pushing and pulling a long spring (slinky).

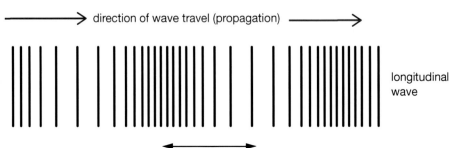

longitudinal wave

vibration directions

- ❏ In a longitudinal wave the **vibrations** or **oscillations** of the particles are **parallel** to the direction of travel of the wave.

- ❏ Longitudinal waves need a medium (material) to travel through, and as they pass through the medium the particles oscillate (vibrate) backwards and forwards.

- ❏ **Sound** is an example of a longitudinal wave.

Transverse waves

- ❏ This type of wave can be shown by moving a long spring (slinky) or a rope from side to side, or by making ripples in water.

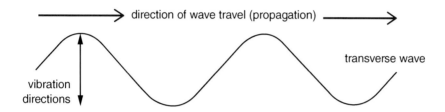

- ❏ In a transverse wave the particles vibrate or oscillate at **right angles** (**perpendicular**) to the direction of travel of the wave.

- ❏ Light, radio and other **electromagnetic waves** are transverse waves in which electric and magnetic fields oscillate at right angles to the direction of travel. Electromagnetic waves can travel through a vacuum.

Reflection and refraction

- ❏ The ripple tank, as shown in the following diagram, is used to produce transverse water waves. It can be used to study wave effects in different situations.

- ❏ A **wavefront** is defined as an imaginary surface joining all points of the wave affected in the same way, e.g. joining all crests or troughs. When plane (straight) waves are produced on the surface of the water in a ripple tank, the wavefronts can be thought of as continuous lines perpendicular to the direction of travel or propagation. This is rather like viewing incoming sea waves from the top of a cliff.

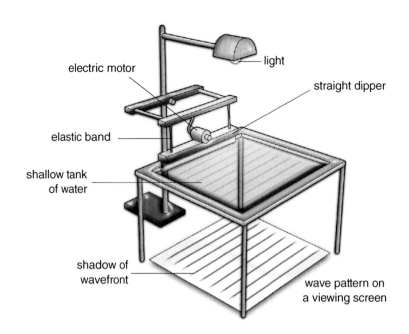

Reflection off a plane surface

❏ A straight dipper can be used to create **plane** waves.

❏ If waves strike a plane barrier at 45° they are reflected as shown.

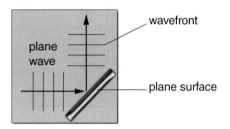

❏ The angle of incidence and the angle of reflection from a plane surface are **equal**.

❏ With reflection, **speed**, **frequency** and **wavelength** remain the **same**. Only the direction changes.

Refraction due to a change in speed

❏ A straight dipper can be used to show **refraction**. By placing a sheet of clear glass in the ripple tank, the water above the glass will be shallower compared to the rest of the ripple tank.

❏ Keeping the frequency of the waves constant, it can be shown that, when a wave moves from one depth into another, it will either **speed up** or **slow down**. Water waves travel faster in deep water, and slower in shallow water. Notice in the following diagram that the wavefronts are closer together in the slower region.

- When a wave travels from deep water to shallow water:
  - wavelength decreases
  - speed decreases
  - frequency stays the same.

  Remember: wave speed = frequency × wavelength

- When a wave moves from one depth to another at an **angle** to to the boundary, it will **change direction**. We say it has been **refracted**.

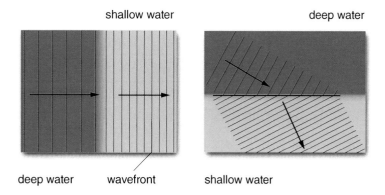

- To help to understand why there is a change of direction when the speed changes, imagine pushing a toy car across some wooden boards towards a region which is covered by sand.

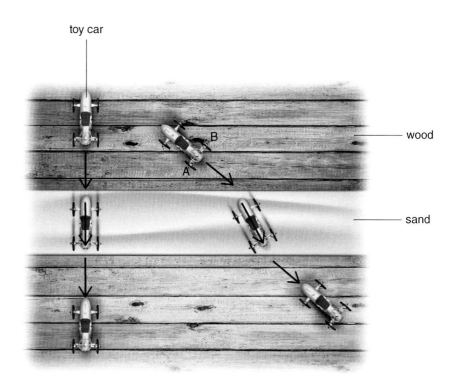

- If the car reaches the sand so that its direction of travel is at an angle of incidence of 0° as shown on the left, it continues in the same direction but slows down because of the increased friction.
- If the car strikes the sand at an angle as shown on the right, wheel A hits the sand first and slows down, but wheel B continues at its original speed. This causes the car to change direction. When wheel B also strikes the sand, the car continues in a straight line in the new direction.
- When the car has crossed the sand, wheel A reaches the wooden board first and speeds up while wheel B continues at the slower speed, until it also reaches the wooden board and speeds up.
- When the car emerges at the other side, it will continue with its original speed and in the original direction.
- All waves, whether transverse or longitudinal, behave as the water waves in the ripple tank, i.e. they can be reflected and refracted.

Example 3b (i)

The frequency (or pitch) of the musical note A is 440 Hz.

(i) State the equation relating wave speed, frequency and wavelength.

(ii) If sound travels at 330 m/s in air, calculate the wavelength of the note A.

Answer

(i) wave speed = frequency × wavelength   or $v = f \times \lambda$

(ii) **Step 1** List all the information in symbol form and change into appropriate and consistent SI units if required.

$f = 440\,\text{Hz}$
$v = 330\,\text{m/s}$
$\lambda = ?$

Step 2 Rearrange the equation.

$v = f \times \lambda \quad \Rightarrow \quad \lambda = \dfrac{v}{f}$

Step 3 Calculate the answer by putting the numbers into the equation.

$$\lambda = \frac{330}{440} = 0.75 \text{m}$$

ALWAYS REMEMBER TO STATE THE UNIT FOR CALCULATED QUANTITIES.

> **Note**
>
>
> Frequency is expressed by
> $$f = \frac{1}{T}$$
> The unit of frequency is the hertz and 1 Hz is a frequency of one wave per second.

Example 3b (ii)

A wave completes 250 complete oscillations in 10s.

(i) State the equation linking frequency and time period.

(ii) Calculate the frequency of the wave.

Answer

(i) frequency = $\frac{1}{\text{time period}}$ or $f = \frac{1}{T}$

(ii) **Step 1** List all the information in symbol form and change into appropriate and consistent SI units if required.

T = time for 1 oscillation = 10 ÷ 250 = 0.04 s
f = ?

Step 2 Calculate the answer by putting the numbers into the equation

$$f = \frac{1}{0.04} = 25 \text{Hz}$$

ALWAYS REMEMBER TO STATE THE UNIT FOR CALCULATED QUANTITIES.

Section 3c The electromagnetic spectrum

- The electromagnetic (e.m.) spectrum is a family of **electromagnetic waves** that travel at the same high speed in free space (vacuum); they have different wavelengths and frequencies.

- Electromagnetic waves don't need anything to travel in, i.e. they can travel through a vacuum.

- Radio waves, microwaves, infrared, visible light, ultraviolet, x-rays and gamma (γ) radiation are all parts of the electromagnetic spectrum.

- The different wavelengths of the different types of wave are shown in the diagram below. Note, for example, that infrared radiation, responsible for the transfer of thermal energy by radiation, has a slightly longer wavelength than visible light.

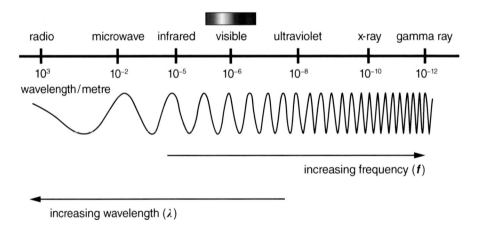

- All electromagnetic waves are **transverse waves** and transfer energy but not matter.

- The **speed** of **all** electromagnetic waves in a vacuum (or approximately in air) is 3.0×10^8 m/s.

The dangers of e.m. radiation

❑ Exposure to large doses of any type of electromagnetic radiation can be harmful, but it is the high frequency (short wavelength) waves which are the most dangerous.

- Microwaves, of the frequency used in microwave ovens, cause internal heating of body tissues and may cause burning.
- Infrared radiation is felt as heat; too much can cause skin burns.
- Ultraviolet radiation occurs naturally in sunlight and it is what causes our skin to turn darker; too much exposure to sunlight can lead to skin cancer. Over-exposure to ultraviolet radiation can cause eye damage and even blindness. Eye protection (e.g. sunglasses) can help to protect the eyes.
- Even small doses of very short wavelength e.m. radiation can be harmful; both x-rays and gamma rays can damage living cells, causing mutations and the death of cells and leading to cancer. To reduce the risk we should limit our exposure to such radiation by limiting the time we are exposed, keeping a safe distance away and using appropriate shielding. This is particularly important for those who work with such radiation (see pages 220-21).

Uses of electromagnetic radiation

❑ Electromagnetic radiation has many uses.

- Radio waves are used for communication and terrestrial radio and TV broadcasts.
- Radio telescopes detect radio waves from space; they have an advantage over optical telescopes because radio waves are not blocked by clouds.

- Microwave radiation is absorbed by water molecules so it is used in microwave ovens for cooking.
- Microwave radiation can be used to transmit signals from mobile phones.
- Some microwave radiation can pass through the Earth's atmosphere and is used for satellite transmission.
- Police speed guns use microwaves because they reflect well from hard surfaces.

- Infrared radiation is used for heating; our bodies detect it as warmth.
- Infrared radiation is an indication of temperature so can be used in thermometers.
- Infrared radiation can also be used as an indicator of abnormal temperatures in parts of the human body which might be caused by illness.
- Infrared tracking cameras can be used to detect movement of animals or humans at night, as the camera responds to the heat of the body.
- Remote control devices send out a beam of infrared which is detected by televisions etc.

- Optical devices use visible light, e.g. cameras.
- Optical fibres are thin glass fibres through which light can be transmitted and reflected, e.g. endoscopes (see total internal reflection pages 109-11).

- Fluorescent lamps convert ultraviolet (UV) light to visible light.
- Ultraviolet light is germicidal and can be used to kill bacteria.
- Ultraviolet light can be used to detect forged banknotes; real notes have security features which are only visible in UV light.

- X-rays can be used to detect bone fractures and other internal abnormalities; they pass through skin and body tissue but not bone.
- X-rays are similarly used in security scans in airports.
- X-rays can detect cracks and fatigue in industrial equipment.

- Gamma rays are used to kill cancer cells, in radioactive tracers used in medical diagnosis, and to sterilise hospital equipment.
- Food can also be sterilised by gamma rays.

Summary of different electromagnetic waves

- ❑ *Gamma (γ-) rays*
    - very short wavelength
    - dangerous (γ-rays can kill living cells)
    - pass through skin and soft body tissue
    - come from radioactive substances such as uranium
    - used to kill cancer cells and sterilise hospital equipment

- ❑ *X-rays*
    - short wavelength
    - dangerous (x-rays can kill living cells)
    - pass through skin and soft body tissue but not bone or metal
    - used to photograph internal structures and in security systems

- ❑ *Ultraviolet rays*
    - wavelength a little shorter than visible light
    - cause tanning and can damage the skin if over-exposed
    - used to check for forged banknotes and in fluorescent lamps

- ❑ *Visible light*
    - mid-range wavelength
    - composed of a spectrum of colours
    - comes from the Sun and other luminous objects
    - used in optical fibres and photography

- ❑ *Infrared rays*
    - wavelength a little longer than visible light
    - come from the Sun and any other hot objects
    - used in TV remote controls, heaters and night photography

- ❑ *Microwaves*
    - long wavelength
    - used in mobile phone networks, microwave ovens and satellite communication

- ❑ *Radio waves*
    - very long wavelength
    - used for terrestrial communications and television and radio broadcasts

Section 3d Light and sound

Reflection of light

- Properties of light:
  - it travels as **transverse** waves
  - it transfers energy
  - it can travel in a vacuum
  - it travels at a speed of 3.0×10^8 m/s in air or a vacuum.

- A light ray is a narrow beam of light that travels in a straight line. A light ray that reflects from a surface such as a **plane mirror** obeys the **law of reflection**:

  The angle of incidence is equal to the angle of reflection, where both angles are measured to the **normal**.

- The **normal** is a construction line at **90°** to the mirror, at the point where the light ray meets the mirror. It allows the angles shown in the diagram to be measured.

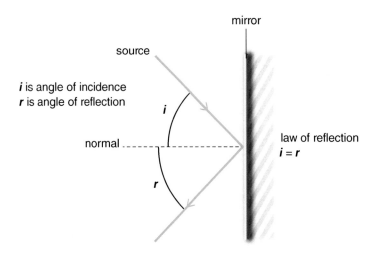

angle of incidence (i) = angle of reflection (r)

> **Note**
>
> During reflection, **speed**, **frequency** and **wavelength** do not change; only **direction** changes.

❑ A plane mirror forms an **image** of an object, which has these properties:
- upright but **laterally inverted**, i.e. the image is reversed left to right
- the same size as the object
- the same distance behind the mirror as the object is in front
- **virtual**, which means the image cannot be formed on a screen.

❑ Plane mirrors are used in periscopes, kaleidoscopes and dressing table mirrors.

Finding the position of the image in a plane mirror

Method

1. Draw any two incident rays from the object to the mirror.
2. Draw in the reflected rays from these incident rays, making sure **angle of incidence = angle of reflection**.
3. Continue these reflected rays back straight behind the mirror using a dotted line to the point where they appear to come from.

❑ The image is formed where the rays appear to come from behind the mirror. This is a **virtual image**. If you placed a screen at position 4 in the diagram below, nothing would show.

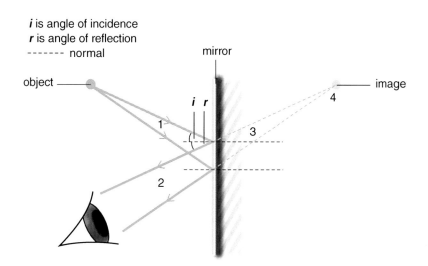

Example 3d (i)

(i) An object is placed in front of a mirror. Construct a ray diagram showing where the position of the image will appear.

(ii) Describe two properties of the image.

Answer

(i) **Step 1** Draw a ray A from the top of the object to the mirror. Mark in the normal and measure, using a protractor, the angle of incidence *i*.

Step 2 Construct the reflected ray such that the angle of reflection *r* is equal to the angle of incidence *i*.

Step 3 Draw a second ray B in the same way.

Step 4 Use dotted lines to show where the reflected rays appear to come from. This will be the top of the image.

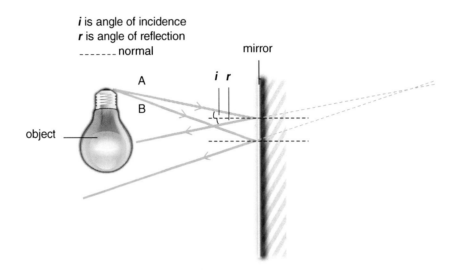

Step 5 Repeat with two rays, C and D, from the bottom of the object.

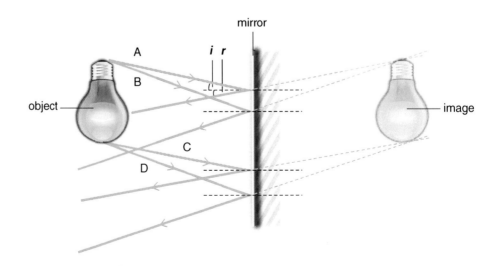

(ii) - The image is the same way up as the object (upright).
- It is the same size as the object.

Refraction of light

❑ Light rays can **change direction** when passing from one material into another because the rays move at **different speeds** in the two materials. This is because light travels as a wave and is refracted (see pages 94-95).

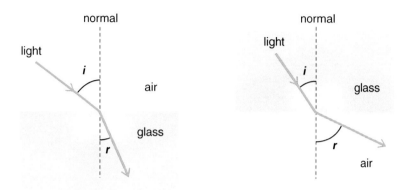

❑ Notice that the **angle of incidence** *i* and the **angle of refraction** *r* always lie between the light ray and the normal.

❑ A material in which light travels slowly is said to be **optically dense**.

❑ When travelling from an optically **less dense** material to a **more dense** material (such as air to glass), light **bends towards** the **normal** (diagram on the left above).

❑ When travelling from an optically **more dense** material to a **less dense** material (such as glass to air), light bends away from the **normal** (diagram on the right above).

❑ When light enters a different material:
   - the frequency stays the same
   - the wavelength changes
   - the speed changes.

> **Note**
>
> When a light ray travels **along the normal** from air into glass, **it passes straight through** (undeviated), but it slows down.

Refractive index

❑ The **refractive index** *n* indicates how the direction of light changes on passing from one material to another. It is one of the few quantities that **does not have a unit**; it is just a number.

❑ Refractive index, the angle of incidence and the angle of refraction are related by the equation:

$$\text{refractive index} = \frac{\text{sine of the angle of incidence}}{\text{sine of the angle of refraction}}$$

$$n = \frac{\sin i}{\sin r}$$

n = refractive index (no unit)
i = angle of incidence (°)
r = angle of refraction (°)

This relationship is **Snell's Law**.

Note

Refractive index is **greater** than 1 (*n* > 1) when light goes from a **less dense** to a **more dense** medium.

For a ray travelling **from air into a material**, *n* can be considered to be the refractive index of the material.

Example 3d (ii)

Light travels from air into a glass block of refractive index 1.4.
The angle of refraction in the glass is 35°.

(i) State the equation linking the refractive index to the angles of incidence and refraction.
(ii) Calculate the angle of incidence in air.

Answer

(i) $\text{refractive index} = \frac{\text{sine of the angle of incidence}}{\text{sine of the angle of refraction}}$ or $n = \frac{\sin i}{\sin r}$

(ii) **Step 1** List all the information in symbol form and change into appropriate and consistent SI units if required.

$n = 1.4$
$r = 35°$
$i = ?$

Step 2 Rearrange the equation.

$$n = \frac{\sin i}{\sin r} \quad \Rightarrow \quad \sin i = n \times \sin r$$

$$i = \sin^{-1}(n \times \sin r)$$

(The term $\sin^{-1}(x)$ means the angle whose sine value is x.)

Step 3 Calculate the answer by putting the numbers into the equation.

$$i = \sin^{-1}(n \times \sin r) = \sin^{-1}(1.4 \times \sin 35)$$
$$i = \sin^{-1}(1.4 \times 0.574) = \sin^{-1} 0.804$$
$$i = 53° \quad \text{(to 2 sig. figs.)}$$

ALWAYS REMEMBER TO STATE THE UNIT FOR CALCULATED QUANTITIES.

Top Tip

When a light ray travels from a **more dense** material (glass) to a **less dense** material (air), the **refractive index must be less than 1** ($n<1$).

Make sure you know which way the light is travelling. If you are given $n>1$ for light travelling from air to glass, but in the question the light is travelling from glass to air, you must first calculate the correct refractive index by finding the **reciprocal**.

If $n_{\text{air to glass}} = 1.43$ then $n_{\text{glass to air}} = \dfrac{1}{1.43} = 0.7$

Once you have the correct refractive index n, the formula above can be used as normal.

Finding the refractive index of glass

❏ The following experiment describes how to measure the angle of incidence and angle of refraction to calculate the refractive index of glass.

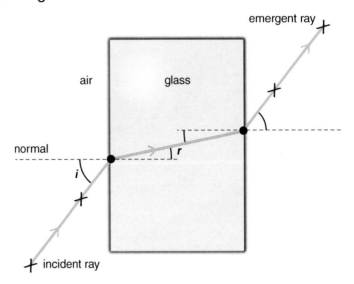

Method

1. Place the glass block in the middle of a plain sheet of paper; trace around the block in pencil.
2. Position a raybox so that the light from it strikes the glass block at an angle.
   Remember: If the light strikes the glass boundary at 90°, it will pass straight through (undeviated).
3. Mark the positions where the light meets the glass boundary and where it leaves the glass boundary with dots (see the diagram).
4. Mark two crosses (or place optical pins) on the paper along the incident ray and the emergent ray at least 5 cm apart.
5. Remove the glass block and switch off the raybox.
6. Using a ruler, complete the lines between the dots and the crosses.
7. Draw in the normal at **90°** to the glass boundary where the light strikes the boundary.
8. Draw the second normal where the light leaves the glass boundary, again at **90°**.
9. Using a protractor, measure the angles of incidence i and refraction r as shown on the diagram above.

10. Repeat steps 2 to 9 for different angles of incidence.
11. Plot a graph of sin *i* against sin *r*, which should be a straight line graph passing through the origin. Check and repeat any anomalous results.
12. The refractive index can be found from the gradient of the graph as:

$$n = \frac{\sin i}{\sin r}$$

Total internal reflection

- ❑ When light travels from a **more dense medium** to a **less dense medium**, e.g. from glass to air, the angle of refraction is greater than the angle of incidence (see diagram on page 105).

- ❑ It follows that, as the angle of incidence increases, the angle of refraction will eventually equal 90°. The angle of incidence for which this occurs is called the **critical angle**.

- ❑ If the angle of incidence is further increased, the light becomes **totally internally reflected**, and obeys the laws of reflection.

- ❑ If a ray of light is shone into a semi-circular glass block along a radius, the light will not change direction as it enters the block, because the angle of incidence is zero. However there will be a change of direction as the light exits the block.

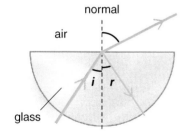
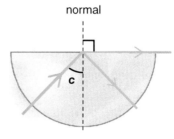
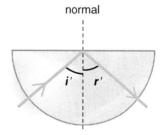

| For small angles of incidence, the ray splits into a bright refracted ray and a dim internally reflected ray. | At the critical angle **c**, most of the light is refracted at 90° along the glass surface; the internally reflected ray has become a little brighter. | For angles of incidence greater than the critical angle, **total internal reflection** occurs. No light is refracted. |

- ❑ The **critical angle** is defined as the angle in the denser material above which total internal reflection occurs.

- ❑ Each material has its own critical angle for light travelling from the material into air. For example, the critical angle of glass is 42° and for water it is 49°.

- ❑ Consequently, light incident at angles greater than 42° for glass and greater than 49° for water will be completely reflected within the material and none of it will be refracted into the air – total internal reflection.

- ❑ The inside surface of diamond, water or glass can therefore act like a mirror depending on the angle at which light strikes it.

- ❑ The critical angle **c** is related to the refractive index **n** by the equation:

$$\text{sine of the critical angle} = \frac{1}{\text{refractive index}}$$

$$\sin c = \frac{1}{n}$$

n = refractive index (no unit)
c = critical angle (°)

Uses of total internal reflection

- ❑ Total internal reflection is used in **fibre optic cables**. A fibre optic cable is made up of a bundle of very thin glass fibres called **optical fibres**. The light travels along the fibre by being repeatedly totally internally reflected inside the glass because the **angles of incidence** are always **greater** than the **critical angle** of the glass.

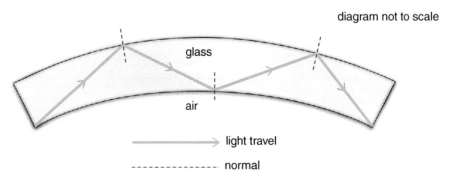

diagram not to scale

→ light travel
------- normal

- Fibre optic cables are used for digital TV transmission, telephone cables, internet communications and in medical devices such as endoscopes. An endoscope consists of a long, thin, flexible tube containing an array of optical fibres. Using a light source and video camera at the end outside a patient's body, images of the inside of the body can be seen on a screen.

- The inside of a glass prism can be used as a mirror. Total internal reflection takes place on the **longest** face of the prism, as shown in the diagram, and occurs because the angle of incidence (45°) is greater than the critical angle (42°). This arrangement is used in **periscopes**.

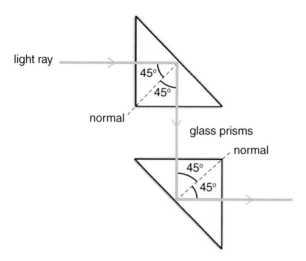

Dispersion of light

- Light of different colours has different wavelengths.

    Red light has a longer wavelength than green light, which has a longer wavelength than blue light.

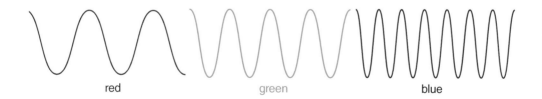

❏ **White** light is made up of a **continuous** range of colours, but can be thought of as **seven** different colours. The colours travel at the same speed of 3×10^8 m/s in free space (vacuum) but slow down when they enter a denser medium like glass. The different wavelengths slow down by different amounts and so refract by different angles in a prism. This is known as **dispersion** and a **spectrum** is produced, as shown in the following diagram.

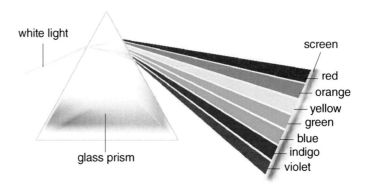

❏ The colours of the **visible spectrum** in order of **decreasing wavelength** can be memorised using the mnemonic: **ROYGBIV**: Richard Of York Gave Battle In Vain.

> **Top Tip**
>
> Always describe white light **splitting up** into a **spectrum of light** and **not** 'colours of the rainbow'.

Sound

❏ Sound travels from a source as waves. The **speed of sound in air** is approximately **330 m/s** at 0 °C. As the temperature increases, the speed increases.

❏ Sound waves are **longitudinal waves**; the particles (e.g. molecules of air) vibrate parallel to the direction of travel of the wave (see pages 92-93).

❏ Sound waves are a series of variations in pressure:
- areas where there is high pressure (where the molecules are squashed together) are known as **compressions**

- areas where there is low pressure (where the molecules are further apart) are known as **rarefactions**.

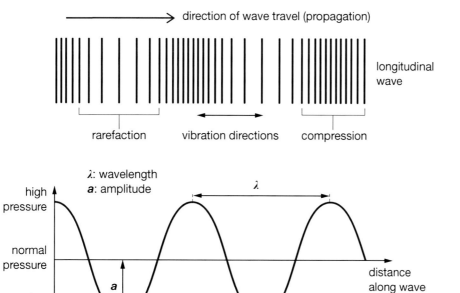

> **Top Tip**
>
>
>
> Sometimes questions about sound can show what looks like a **transverse wave**; be careful, as it is a **graphical** representation of **pressure** in a sound wave. Notice in the diagram above how the peaks of the graph correspond to the positions of compression in the longitudinal wave.

❑ Sound waves can be reflected and refracted.

❑ Sound **cannot** travel through a **vacuum**. (There are no particles to vibrate.)

○ Sound is caused by **vibrations**. It travels as a wave through a medium (which may be **solid**, **liquid** or **gas**) and transfers energy.

○ Examples of how sound is produced include:
  - a hammer hitting a nail – the hammer and the nail vibrate, which in turn cause air molecules to vibrate, creating a sound wave that travels to the ear
  - a door slamming – the door and its frame vibrate, causing air molecules to vibrate, creating a sound wave that travels to the ear

- a cricket bat hitting a ball – the bat and ball vibrate, causing air molecules to vibrate, creating a sound wave that travels to the ear
- musical instruments which employ a variety of techniques such as plucking or bowing a string, banging a drum, blowing air into a pipe, creating a sound wave in the air which travels to the ear.

- **Remember:** It is the energy that is transferred, not the matter. In air, a sound wave makes the air molecules vibrate backwards and forwards, and it is only the compressions and rarefactions that travel through the air from the vibrating object to our ears.

- The frequency of the sound wave is the same as the frequency of the vibrating object that makes it.

- The sensation of frequency is often referred to as **pitch**. A high-frequency sound is heard by the human ear as a high-pitched sound.

- Sound waves have a wide range of wavelengths and, therefore, frequencies. Our ears, however, do not respond to all frequencies. The normal **human hearing** frequency range is 20 Hz – 20 000 Hz. It is usual for this range to decrease as we get older.

- Sound above 20 000 Hz is known as **ultrasound**. In medicine an ultrasound scan (sometimes called a sonogram) is a painless test that creates images of organs and structures inside your body using reflection of ultrasound waves. Ultrasound scans are widely used on pregnant women, to check the health and development of their unborn babies.

- Ultrasound can also be used to test for flaws in machines, where cracks in structures will reflect the ultrasound waves and so can be detected.

- Sonar systems in ships use ultrasound for navigation and to detect and locate objects under the surface of the sea.

- Sound with a frequency lower than 20 Hz is called infrasound. Infrasound is used by some large animals for communication. For example, whales, giraffes and elephants can communicate over many miles using infrasound.

Doppler effect

- When the source of waves, such as sound waves, moves towards or away from an observer, there is a change in its observed frequency. If you hear a police car or an ambulance with its siren sounding, you will notice that when the source of sound is coming towards you it has a higher frequency (pitch) than when it is going away from you. This effect is called the **Doppler effect**, and the change in frequency is called the **Doppler shift**.

- The diagram above shows a motor cycle travelling from left to right. The lines show the wavefronts as the sound of the motor cycle travels through the air. For the boy the wavefronts are pushed together more because the motor cycle is travelling towards him and the sound he hears has a shorter wavelength, or a higher frequency, than the sound heard by the girl.

- All waves demonstrate the Doppler effect when the source is moving relative (towards or away from) the observer, but the effect is less obvious when the speed of the source is very small compared with the speed of the wave.

❏ Light travels at 300 000 000 m/s (3×10^8 m/s) in air, so we rarely notice the Doppler effect in light, but it is seen when viewing stars and distant galaxies. The light from distant galaxies that are moving away from Earth undergoes a **'red shift'**.

We saw earlier in this section (page 112) that white light is composed of a spectrum of colours with varying wavelengths. A shift to the red end of the spectrum in the light from a galaxy means the **wavelength has increased** (and the frequency has decreased), so it must be moving away from us (see pages 257-58).

❏ An observer will notice the same effect when hearing a passing racing car, police siren, train or other source of sound.

Speed of sound

○ The **speed of sound** varies from material to material. A sound wave compresses (pushes together) and rarefies (spreads apart) the particles of the material, and how easily this happens affects the speed of the wave.

○ Sound travels fastest in solids because the molecules in a solid are very close together.

○ Sound travels second fastest in liquids because the molecules in a liquid are only slightly further apart.

○ Sound travels slowest in gases because the molecules in a gas are far apart.

○ **Remember:** Sound cannot travel through a vacuum because there are no particles.

○ Some examples of the speed of sound in different media are given below.
- In dry air at 0°C the speed is 330 m/s.
- In dry air at 20°C the speed is 340 m/s.
- In water the speed is 1500 m/s.
- In gold the speed is 3200 m/s.
- In steel the speed is 5800 m/s.

○ The following experiment describes how to determine the speed of sound in air.

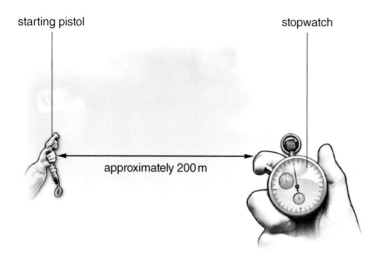

Method

1. A student with a stopwatch stands **a sufficiently long distance** from a teacher who has something that makes a loud noise, e.g. a starting pistol or two blocks of wood that can be banged together.

2. The distance apart in metres is measured.

3. The teacher fires the pistol in the air, or bangs the wooden blocks together.

4. The student starts the stopwatch when he/she sees the puff of smoke or the blocks being banged together and stops it when he/she hears the bang.

○ The speed of light is very large and it takes a very small fraction of a second for the light to travel from the teacher to the student, so the student is actually measuring the time taken for the sound to arrive.

Calculations to be made
Use the formula to calculate the speed of sound:

$$\text{average speed} = \frac{\text{distance moved}}{\text{time taken}} \quad \text{or} \quad v = \frac{s}{t}$$

Improvements to the experiment

The experiment could be improved by:

1. repeating it to find a mean value, and using different students to take the measurements

2. increasing the distance between the pistol and the stopwatch (but not so far that the sound cannot be heard), so that the timekeeper's reaction time has less effect on the time

3. swapping the position of the pistol and the stopwatch to take account of any wind direction, because the direction of the wind will affect the speed of the sound waves.

> **Note**
>
>
> The speed of light (3.0×10^8 m/s) is very much greater than the speed of sound (330 m/s). This is why we see lightning first and then hear thunder a little later.

Example 3d (iii)

Priya sees a flash of lightning and begins to count 'one thousand, two thousand, ...' until she hears the thunder. The time it takes to say 'one thousand' is about one second, so she is able to approximate the time between the lightning flash and the thunder.

The speed of light is very large and it takes a very small fraction of a second for the light to travel to Priya, so she is actually measuring the time taken for the sound to arrive.

(i) State the equation linking the speed of sound with the distance moved and the time taken.

(ii) If the speed of sound is given as 340 m/s and the time estimated by Priya is 5 seconds, how far away is the storm?

Answer

(i) average speed = $\dfrac{\text{distance moved}}{\text{time taken}}$ or $v = \dfrac{s}{t}$

(ii) **Step 1** List all the information in symbol form and change into appropriate and consistent SI units if required.

$v = 340 \, m/s$
$t = 5 \, s$
$s = ?$

Step 2 Rearrange the equation.

$$v = \frac{s}{t} \quad \Rightarrow \quad s = v \times t$$

Step 3 Calculate the answer by putting the numbers into the equation

$s = 340 \times 5 = 1700 \, m$

ALWAYS REMEMBER TO STATE THE UNIT FOR CALCULATED QUANTITIES.

The speed of sound and echoes

- **Remember:** Sound is a longitudinal wave that can travel to a surface, be reflected and travel back again as an **echo**.

- If the distance between the source of sound and the reflective surface, and the time taken are known, then the speed of the sound can be calculated using $v = s/t$.

Example 3d (iv)

Sara stands 200 m from a large wall and claps together two large blocks of wood. She hears an echo. She claps regularly and times her claps so that they coincide with the echoes. A friend standing beside her with a stopwatch records the time taken for the sound of 10 claps to reach the wall and echo back again as 12.4 s.

(i) State the equation linking speed of sound to distance moved and time taken.
(ii) Calculate the speed of sound.

Answer

(i) $\text{average speed} = \dfrac{\text{distance moved}}{\text{time taken}} \quad \text{or} \quad v = \dfrac{s}{t}$

(ii) **Step 1** List all the information in symbol form and change into appropriate and consistent SI units if required.

$s = 2 \times 200\,\text{m}$ (the distance to the wall and back)
The time taken for sound to travel there and back for 10 claps is 12.4 s, so
$t = 12.4 \div 10 = 1.24\,\text{s}$
$v = ?$

Step 2 Calculate the answer by putting the numbers into the equation.

$$v = \frac{2 \times 200}{1.24} = 322.58 = 320\,\text{m/s} \quad \text{(to 2 sig. figs)}$$

ALWAYS REMEMBER TO STATE UNITS OF CALCULATED QUANTITIES.

○ If the speed of sound in a medium and the time taken are known, then the distance the wave has moved can be calculated using $s = v \times t$.

Example 3d (v)

A fishing boat uses sonar (sound navigation and ranging) to detect shoals of fish under the water. A pulse of sound is sent from the surface, and is reflected back after it hits the shoal of fish. Sound travels at 1500 m/s in water and the time between the ping and its echo is 0.40 s.

(i) State the equation linking speed of sound to distance moved and time taken.
(ii) Calculate the depth of the water where the shoal is located.

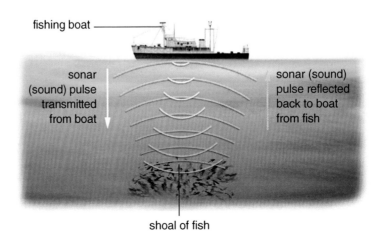

Answer

(i) average speed = $\dfrac{\text{distance moved}}{\text{time taken}}$ or $v = \dfrac{s}{t}$

(ii) **Step 1** List all the information in symbol form and change into appropriate and consistent SI units if required.

$v = 1500\,\text{m/s}$
The time taken for the sound to travel there and back is 0.40s, so
$t = 0.40 \div 2 = 0.20\,\text{s}$
$s = ?$

Step 2 Rearrange the equation.

$v = \dfrac{s}{t} \quad \Rightarrow \quad s = v \times t$

Step 3 Calculate the answer by putting the numbers into the equation.

$s = 1500 \times 0.20 = 300\,\text{m}$

ALWAYS REMEMBER TO STATE THE UNIT FOR CALCULATED QUANTITIES.

Sound wave characteristics

- Frequency, pitch, wavelength and amplitude are terms associated with sound waves.

- **Frequency** is the number of complete oscillations, or wavelengths, produced per second and its unit is the **hertz** (Hz).

- **Pitch** is how **high** or **low** a sound seems to our ears, and is dependent on frequency.

- **Wavelength** is the distance between any point on a wave and the next similar point on the wave. (It is often measured from crest to crest, but it doesn't matter as long as it is between two similar points.) Its unit is the **metre** (m).

- **Amplitude** is the maximum disturbance of the wave particles from their equilibrium position. It determines how **loud** or **quiet** a sound is – the greater the amplitude, the louder the sound.

Oscilloscope wave traces

- If a microphone is connected to an oscilloscope, the sound received by the microphone is transferred electrically and the oscilloscope trace can be used to compare frequency (pitch) and amplitude (loudness).

- The diagrams show graphical representations of sound waves as seen on an oscilloscope screen.

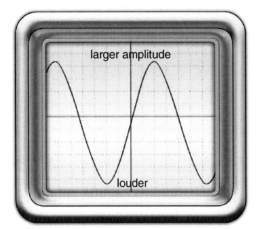 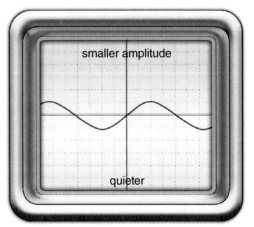

- Notice that the frequencies of the loud sound and the quiet sound are the same in the traces above. Only the amplitude is different.

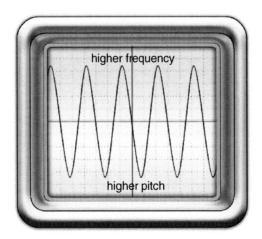 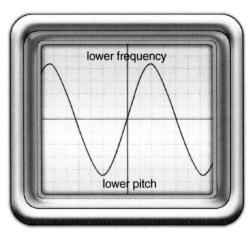

- Notice that the amplitudes are the same for the high- and low-pitch sounds in the lower two diagrams. Only the frequency differs.

- **Remember:** Sound waves are longitudinal waves and not transverse waves. The diagrams above are graphical representations of the pressure variations.

Section 4   Energy resources and energy transfers

Section 4a   Units

❏ In this section you will come across the following units:
   - the **kilogram** (kg) is the unit of mass
   - the **joule** (J) is the unit of energy
   - the **metre** (m) is the unit of length
   - the **second** (s) is the unit of time
   - the **metre per second** (m/s) is the unit of speed or velocity
   - the **metre per second squared** (m/s^2) is the unit of acceleration
   - the **newton** (N) is the unit of force
   - the **watt** (W) is the unit of power.

Section 4b   Energy transfers

❏ Energy can be **stored** or **transferred**. Energy **cannot** be **created** or **destroyed**. We say that energy is **conserved**. This is the **principle of conservation of energy**. In other words:

total energy into system = total energy out of system

❏ The unit of energy is the **joule** (J).

❏ There are various forms of **energy store**. Some examples are given below:
   - **elastic** (or **strain**) – stored energy when an elastic material is stretched or squashed
   - **kinetic** – the energy a body has because of its movement
   - **gravitational potential** – energy stored due to a change in height (for example, water at the top of a waterfall has gravitational potential energy because it has the potential to fall and to transfer into other forms of energy)
   - **thermal** – energy stored due to the internal energy (vibration/movement) of atoms and molecules, i.e. heat
   - **chemical** – energy stored in fuels such as petrol and food and released when the fuel is combined with oxygen

- **nuclear** – energy stored inside the nucleus of an atom and released by fission, fusion or radioactive decay
- **magnetic** – energy stored in two magnets that are either attracting or repelling each other
- **electrostatic** – energy stored in two charges that are either attracting or repelling each other.

❏ Energy can be transferred from one store to another. This can be:
- **mechanically** – when a force moves through a distance
- **electrically** – when charges move as in an electric current
- **by heating** – energy is transferred when there is a temperature difference
- **by radiation** – energy can be transferred by electromagnetic radiation such as light and infrared and by sound.

❏ Some examples of how energy is transferred from one store to another are given below:

1. The chemical energy in a battery is transferred electrically to a lamp which then transfers energy to the surroundings by radiation (light) and by heating (thermal energy).

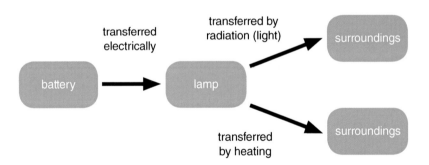

2. A bicycle at the top of a hill has a store of gravitational potential energy. If the cyclist freewheels down the hill, the gravitational potential energy is transferred mechanically and the bicycle gains kinetic energy. Some energy is also transferred by heating to the surroundings.

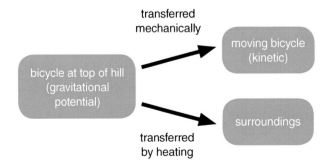

3. A catapult slinging a stone transfers the elastic stored energy of the catapult mechanically to the stone, which gains kinetic energy. Some of the catapult's energy is also transferred by heating to the surroundings and by radiation as sound.

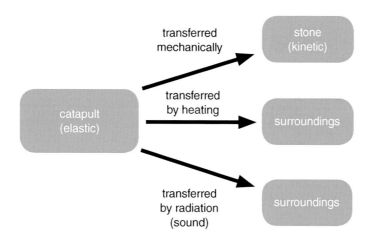

- In all changes, some energy is always transferred by heating to the surroundings as thermal energy (often due to friction) and is often wasted.

- **Efficiency** is the term used to describe how good a device is at transferring one form of energy into another. It is better for a car to be 70% rather than 50% efficient. An efficiency of 70% means the car is transferring seven-tenths of the chemical energy from the fuel into useful energy. The other 30% or three-tenths is transferred, to the surroundings either thermally or by radiation as sound, both of which are less useful.
  N.B. In cold countries some of this 30% would be useful energy rather than wasted as it would keep us warm in the car.

- **Useful energy** can be described as the type of energy we want from a device or what it's built (designed) to do. For example, a television set transfers energy to the surroundings as light, thermal energy and sound. We want to see and hear the television, so in this example light and sound are useful. We do not need the thermal energy; therefore it is described as 'wasted'.

  The equation for efficiency is:

  $$\text{efficiency} = \frac{\text{useful energy output}}{\text{total energy input}} \times 100\%$$

❏ For example, 100J of chemical energy from a battery is transferred electrically into 100J of output energy in a flashlight. A flashlight is used to provide light.

If 15J comes out as light, then the other 85J is wasted as thermal energy, and the efficiency of the flashlight is 15%.

If 10J comes out as light, then 90J is wasted as thermal energy, and the efficiency of the flashlight in this case is 10%.

100J of **input energy** is **always** transferred into **100J** of **output energy**.

> *Top Tip*
>
>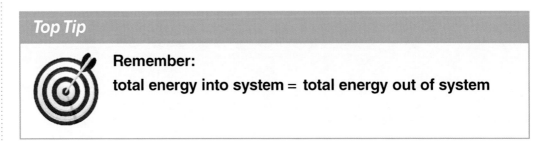
>
> **Remember:**
> **total energy into system = total energy out of system**

❏ A **Sankey diagram** is a useful visual representation showing energy input and output.

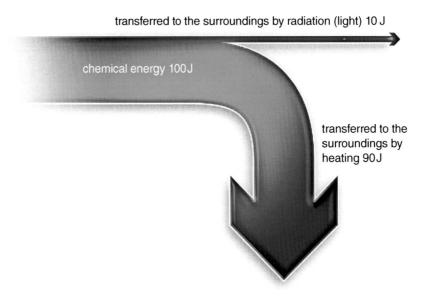

The diagram above represents the energy of a light bulb with an efficiency of 10%, i.e. of the 100J input only 10J of the output is transferred by radiation (light). The rest is 'wasted' as the surroundings are heated.

When drawing a Sankey diagram, make sure that the thickness of the arrows represents the relative amounts of energy.

❑ Sankey diagrams can also be used to represent more complex examples of energy transfer.

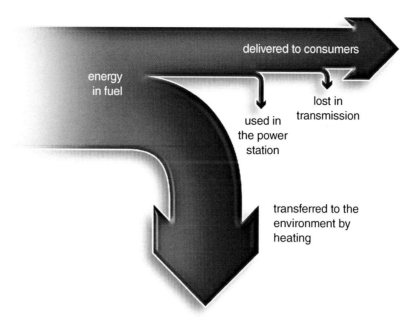

The diagram above shows how the energy of the fuel in a power station is used. Only about one-third of the input energy is delivered to consumers.

Thermal energy transfer

❑ The three principal methods of thermal energy transfer are:
1. conduction
2. convection
3. radiation.

❑ **Conduction** – Particles **gain energy** when heated. In a solid the particles cannot change positions but they vibrate, and this transfers the energy through the solid from particle to particle. In addition metals are particularly good conductors because they have **free-moving electrons**. Collisions between the vibrating particles and the moving electrons cause energy to be transferred quickly.

- **Convection** – A liquid or gas (fluid) **expands** as it is heated and it becomes **less dense**; this causes the hot fluid to **rise** and the cooler fluid above it to **fall** in a circular fashion. The rising of the hot gas or liquid sets up a **convection current**.

  Remember:
  - It is **NOT** 'heat' that rises; it is the molecules of hot liquid or gas that rise.
  - It is not the particles themselves that expand, but the spacing between the particles.

- **Radiation** – Energy in the form of **infrared radiation** (part of the electromagnetic spectrum) travels in all directions from any **hot body**. This is how thermal energy is transferred through a vacuum, such as from the Sun through space to Earth.

- The amount of infrared radiated from a body depends on its temperature; a very hot body radiates more infrared than a cool body. A body with a large surface area will radiate more than a body of small surface area at the same temperature.

- Because infrared radiation is an electromagnetic wave it can be reflected. This means that a highly polished surface will reflect radiation away, e.g. a polished metal plate behind the element of an electric fire reflects the thermal energy into the room.

> **Note**
>
>
>
> **Conduction** is the transfer of energy by vibrating particles in **solids**. Good conductors are mostly metals, which have free-moving electrons to carry the energy.
>
> **Convection** is the transfer of energy by **hot gas or liquid rising** and **cold gas or liquid falling**.
>
> **Radiation** is the transfer of energy as an electromagnetic wave. Radiation is the only way thermal energy can travel through a vacuum.

Experiments demonstrating thermal energy transfer

- **Conduction**

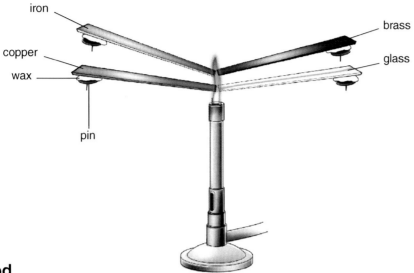

Method

1. Select four strips of different material of equal length.
2. Attach a drawing pin to the end of each strip, using petroleum jelly or candle wax.
3. Heat the other ends of the strips equally.

- The strip that allows the pin to drop first is the best conductor, and so on.
- Glass is a bad conductor, so the drawing pin will stay attached.

- **Convection**

Method

1. Fill the container (see diagram) with cold water.
2. Carefully drop a few crystals of potassium manganate (VII) into the container.
3. Heat gently with a small flame as shown.

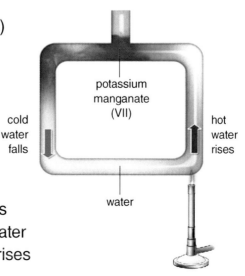

- The purplish pink colour moves in a circular path until all the water becomes coloured. Hot water rises and cold water falls.

❏ *Radiation*

Method

1. The diagram below shows two aluminium plates placed an equal distance from an electric heater. One of the plates is painted matt black and the other has a shiny polished surface. Attach a cork to the back of each plate with wax and turn on the heater.

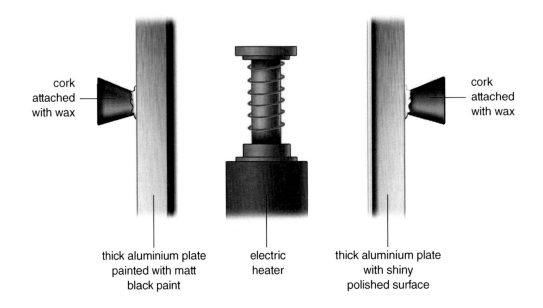

cork attached with wax

cork attached with wax

thick aluminium plate painted with matt black paint

electric heater

thick aluminium plate with shiny polished surface

❏ The cork attached to the matt black plate will drop off first, indicating that the matt black surface is a better **absorber** of thermal radiation than the shiny polished surface. Shiny surfaces are good reflectors of thermal radiation.

❏ The cube shown below is known as Leslie's cube. It has one shiny white surface, one dull white surface, one shiny black surface and one dull black surface.

rubber stopper

Method

1. Fill the cube with boiling water and seal it with a stopper. All four sides of the cube are at the same temperature, because they are all in contact with the boiling water.

2. Place four thermometers or infrared detectors a small distance from each of the surfaces. These will detect how much thermal energy is radiated from each surface. Make sure the detectors are the same distance from each surface to ensure the experiment is fair.

❑ The detector or thermometer at the dull black side will have the highest reading, followed by shiny black, dull white and shiny white. Therefore the dull black surface is the best **emitter** (i.e. the best surface for radiating thermal energy), and the shiny white surface is the worst emitter.

❑ The table below gives a summary of the behaviour of surfaces.

	Best surface	Worst surface
Infrared emitters	dull black	shiny white
Infrared reflectors	shiny white	dull black
Infrared absorbers	dull black	shiny white

Consequences of energy transfer

❑ Some everyday examples of thermal energy transfer are given below.

- We feel the thermal energy radiated by the Sun, electric fires and electric lamps when our skin absorbs the radiation.
- White clothes are often worn in warm weather because they reflect infrared radiation better.
- Highly polished teapots are not good radiators and so keep their contents warmer for longer than black teapots.
- White buildings keep cooler in warm weather than dark ones because they reflect the radiation from the Sun.
- If you heat a metal saucepan of water on top of the cooker, the thermal energy transfers to the water by conduction through the saucepan base; you can see the water moving as convection currents begin to transfer the heat through the water.
- The handle of a saucepan is often made of a poor conductor of thermal energy such as wood or plastic so you don't burn your hand when lifting the pan.
- A cup of hot liquid will stay warmer in a polystyrene cup than in a metal one. In fact, the process of cooling of a liquid is quite complex, but the biggest difference here is that the polystyrene is a poor conductor of thermal energy and the metal is a good conductor, so metal is better at transferring the thermal energy to the surroundings.

Top Tip

Any hot object will cool down more quickly if the **temperature difference** between the object and its surroundings is greater. For example, a cup of hot chocolate will cool down more quickly if the surrounding room is much colder than the drink. A smaller difference in temperature will result in a longer cooling time.

Reducing heat loss from the home

❑ When a house is heated in cold weather, there is thermal energy transfer from the inside of the house to the surroundings. The diagram shows typical values for the proportions of the total thermal energy lost through different parts of a house. (Of course the energy is not really 'lost'; it escapes from the house into the environment.) Thermal energy loss at any given time depends on the difference in temperature between the inside and the outside of the house. If the temperature difference increases, the rate of thermal energy loss increases.

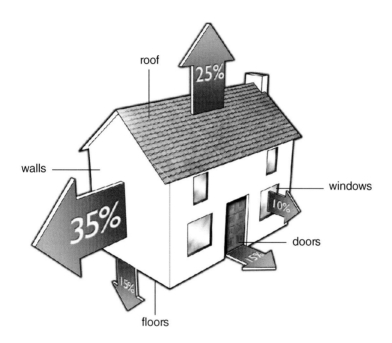

❑ **Loft insulation** – Fibreglass reduces thermal energy loss by conduction, as it is a good **insulator**. This means it is a poor thermal conductor. Fibreglass also prevents thermal energy loss by convection currents, as the fibres trap the air and stop it rising.

❑ **Double glazing** – Air (or other gas) trapped between the sheets of glass reduces convection and conduction. Gases are poor conductors. Radiation will pass through unless the glass has a special reflective coating.

❑ **Floor insulation** – Carpets reduce heat loss by conduction, as the carpet fibres trap air. Air is a very good insulator.

❑ **Wall insulation** – Cavity walls (two layers of bricks with a gap between) can be filled with foam to prevent convection currents in the cavity as well as conduction through the walls.

The domestic radiator

❏ A radiator radiates some infrared – so, if you stand close enough, you can feel the radiation emitted by the surface of the radiator.

However, a radiator is really misnamed since most of the thermal energy is taken away by the hot air that rises from the radiator.

Colder air from the room flows in to replace this hot air, and a **convection** current is formed as shown below. In time this spreads the thermal energy around the whole room.

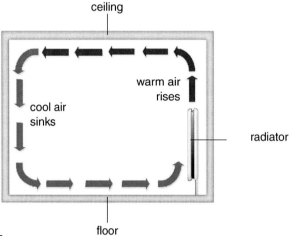

The vacuum flask

❏ A vacuum flask reduces conduction, convection and radiation. When a flask contains a hot liquid, the vacuum between the silvered surfaces (see the diagram) stops energy transfer by conduction and convection.

Silvered surfaces reduce thermal energy loss by infrared radiation, by reflecting the radiation back inside the flask to keep the liquid warm.

The cork at the top and the insulated supports at the bottom reduce energy transfer by conduction. The cork at the top also reduces thermal energy loss by convection and by evaporation of the hot liquid.

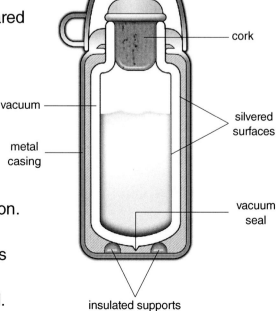

Section 4c Work and power

Work

- **Work** is done when a force is exerted through a distance. More work is done when:
    - the **force** is **larger**
    - the **distance** moved is **greater**.

- The work done is a measure of the amount of **energy transferred** by the force; it has the symbol **W** and, like all forms of energy, its unit is the **joule** (J). For example, if a toy car is pushed through a certain distance it will be given energy – kinetic energy (see below). The work done is equal to the change in kinetic energy. This is a consequence of the principle of conservation of energy.

- **Work done** is related to the **force** and the **distance moved in the direction of the force** by the equation:

    work done = force × distance moved in the direction of the force

    $$W = F \times d$$

    But work done is equal to the energy transferred so:

    $$W = F \times d = E$$

    W = work done (J)
    F = force (N)
    d = distance moved in the direction of the force (m)
    E = energy transferred (J)

Example 4c (i)

A 20 N horizontal force pushes a toy tractor and 50 J of energy is transferred.

(i) State the equation linking energy transferred with force and distance moved.

(ii) Calculate the distance moved.

4 Energy resources and energy transfers

Answer

(i) energy transferred = force × distance moved   or   **E = F × d**

(ii) **Step 1** List all the information in symbol form and change into appropriate and consistent SI units if required.

E = energy transferred = 50 J
F = 20 N
d = ?

Step 2 Rearrange the equation

$$W = E = F \times d \quad \Rightarrow \quad d = \frac{E}{F}$$

Step 3 Calculate the answer by putting the numbers into the equation.

$$d = \frac{50}{20} = 2.5 \text{ m}$$

ALWAYS REMEMBER TO STATE THE UNIT FOR CALCULATED QUANTITIES.

Kinetic energy

❑ **Kinetic energy** is the energy associated with **movement** and is abbreviated as **KE**. Its unit is the **joule** (J).

❑ Mass has the symbol **m**, and its unit is the **kilogram** (kg).

❑ Both **speed** and **velocity** have the symbol **v**, and their unit is the **metre per second** (m/s). When speed or velocity increases, **KE** increases.

Remember: The **magnitude** (size) of the velocity is equal to the speed of the object.

❑ These quantities are related by the equation:

kinetic energy = ½ × mass × speed²

$$KE = \frac{1}{2} \times m \times v^2$$

KE = kinetic energy (J)
m = mass (kg)
v = speed (m/s)

To calculate the speed when you know the kinetic energy and mass, rearrange the above equation to give:

$$v = \sqrt{\left(\frac{2 \times KE}{m}\right)}$$

> **Top Tip**
>
>
> When a car travels at a **constant speed**, its **kinetic energy remains constant** as its speed *v* is not changing. Although its *KE* is not changing, the car still needs fuel (chemical energy) to keep moving at a constant speed because it is doing work against frictional forces.

Example 4c (ii)

A toy car has a mass of 500 g and travels at 60 cm/s.

(i) State the equation linking mass, speed and kinetic energy.
(ii) Calculate the kinetic energy of the car.

Answer

(i) kinetic energy = ½ × mass × speed² or $KE = \frac{1}{2} \times m \times v^2$

(ii) **Step 1** List all the information in symbol form and change into appropriate and consistent SI units if required.

v = 60 cm/s = 0.60 m/s
m = 500 g = 0.500 kg
KE = ?

Step 2 Calculate the answer by putting the numbers into the equation.

$$KE = \frac{1}{2} \times 0.500 \times 0.60^2 = 0.09 \text{ J}$$

Remember: It is only speed that is squared.

ALWAYS REMEMBER TO STATE THE UNIT FOR CALCULATED QUANTITIES.

Gravitational potential energy

- When an object is **lifted**, the work done against the force of gravity is transferred into **gravitational potential energy**. It is abbreviated as *GPE* and its unit is the **joule** (J).

- The change in gravitational potential energy is equal to the work done by or against the force of gravity.

- The change in gravitational potential energy *GPE* when an object moves through a height can be expressed by the following equation:

$$\text{gravitational potential energy} = \text{mass} \times \text{gravitational field strength} \times \text{height}$$

GPE = m × g × h

GPE = gravitational potential energy (J)
m = mass (kg)
g = gravitational field strength (N/kg)
h = height (m)

Remember: *g* is sometimes referred to as the acceleration of free fall, which has the same numerical value but is measured in m/s² (see page 31).

- Energy is conserved. If we know the energy of an object at any one point, then calculating the energy at any other point is possible, ignoring energy transferred by heating.

Top Tip

You could be asked to find the speed of an object dropped from a height. The key here is to understand that the *GPE* at the **top** is equal to the *KE* at the **bottom** just before it hits the ground.

At the top
The object has no *KE* since it is initially at rest (stationary). It has maximum *GPE* since it is high up.

Near the bottom
When the object is at the bottom, it has maximum *KE* since it gets faster as it falls. All its *GPE* has been transferred to *KE* if air resistance is negligible.

Example 4c (iii)

A boy has a mass of 75 kg and dives from a diving board 8.0 m above the surface of the water.

(i) State the equation linking **GPE** with mass, the gravitational field strength and height.

(ii) Calculate his **GPE** at A (see diagram), relative to that at B, just before he dives.

(iii) Find his **KE** at B just before he hits the water (ignoring air resistance).

(iv) State the equation linking **KE** with mass and speed.

(v) Calculate the boy's speed at B.

(vi) What energy changes occur after he enters the water?

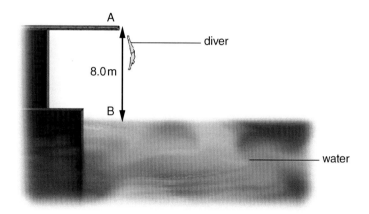

Answer

(i) gravitational potential energy = mass × gravitational field strength × height

or **GPE = m × g × h**

(ii) **Step 1** List all the information in symbol form and change into appropriate and consistent SI units if required.

m = 75 kg
h = 8.0 m
g = 10 N/kg = 10 m/s^2

Step 2 Calculate the answer by putting the numbers into the equation.

GPE = 75 × 10 × 8.0 = 6000 J

(iii) As he falls, his **GPE** is converted to **KE** since energy is conserved.

KE gained = **GPE** lost = 6000 J

(iv) kinetic energy = ½ × mass × speed² or $KE = \frac{1}{2} \times m \times v^2$

(v) **Step 1** Rearrange the equation.

$$KE = \frac{1}{2} \times m \times v^2 \quad \Rightarrow \quad v = \sqrt{\left(\frac{2 \times KE}{m}\right)}$$

Step 2 Calculate the answer by putting the numbers into the equation.

$$v = \sqrt{\left(\frac{2 \times 6000}{75}\right)} = 12.649 = 13\,\text{m/s} \quad \text{(to 2 sig. figs)}$$

ALWAYS REMEMBER TO STATE THE UNIT FOR CALCULATED QUANTITIES.

(vi) When the boy enters the water he slows down because of the resistance of the water, and so loses **KE**. Energy is transferred by heating to the water. Some water may splash up, gaining **KE** and **GPE**. Some energy is transferred as sound.

Top Tip

When an object falls freely in a vacuum under gravity, its gravitational potential energy decreases (since it is getting lower) and its kinetic energy increases (since it is getting faster) by the same amount. The **total energy** (**GPE** + **KE**) stays the **same** since energy is conserved.

In air, some of the energy will be transferred by heating because of work done against air resistance.

Top Tip

If a runner runs up a hill at constant speed, her **KE** remains the same because the speed is **constant**, and her **GPE** increases because she is getting **higher**. Her stored **chemical energy** decreases as she runs and some of the energy is transferred by heating. The **total energy** will remain the **same** since energy is not lost: it just changes from one form into another.

Determining speed from height dropped

❑ For a falling object, in the absence of air resistance:

GPE lost = **KE** gained

$$m \times g \times h = \frac{1}{2} \times m \times v^2 \text{ where } h \text{ is the change in height}$$

The mass **m** can be cancelled from both sides:

$$g \times h = \frac{1}{2} \times v^2$$

This can rewritten as: $v = \sqrt{2 \times g \times h}$

❑ The speed or the change in height can be calculated provided we know one of them, as we know that on Earth the gravitational field strength is 10 N/kg. Consequently, we can calculate the final speed or the initial height **without knowing the mass**.

Example 4c (iv)

A boy drops a golf ball from rest off a cliff, which is 90 m high.

(i) State the equation linking **KE** to mass and velocity.
(ii) State the equation linking **GPE** to mass, **g** and height.
(iii) Calculate how fast the ball is travelling just before it hits the ground.
(iv) Explain why the actual speed in reality is lower than the value calculated in (iii).

Answer

(i) $KE = \frac{1}{2} \times m \times v^2$

(ii) $GPE = m \times g \times h$

(iii) **Step 1** List all the information in symbol form and change into appropriate and consistent SI units if required.

$h = 90 \text{ m}$
$g = 10 \text{ N/kg} = 10 \text{ m/s}^2$
$v = ?$

Step 2 Energy is conserved so

$$m \times g \times h = \frac{1}{2} \times m \times v^2$$

and so $v = \sqrt{2 \times g \times h}$

Step 3 Calculate the answer by putting the numbers into the equation.

$$v = \sqrt{2 \times 10 \times 90} = \sqrt{1800} = 42.42 = 42\,\text{m/s}$$
$$\text{(to 2 sig. figs)}$$

ALWAYS REMEMBER TO STATE THE UNIT FOR CALCULATED QUANTITIES.

(iv) The calculation ignores air resistance and assumes that **no energy** is transferred to the surroundings and that **all the GPE** is transferred into **KE**. In fact, some of the energy will be transferred to the surroundings by heating.

❑ Some energy is **always** transferred by heating (except in a perfect vacuum); therefore the speed will always be lower than the one calculated, as in the previous example.

Example 4c (v)

A skateboarder with a mass of 45 kg rolls along a level path at a speed of 4.6 m/s. Ahead of him is a ramp which is 1.3 m high.

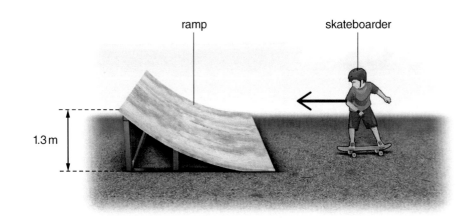

(i) State the equation linking **KE**, mass and speed.
(ii) Calculate the skateboarder's **KE** at the base of the ramp.
(iii) State the equation linking **GPE**, mass, gravitational field strength and height.
(iv) Calculate how high he can travel and state whether he will reach the top of the ramp.
Assume **g** = 10 N/kg. Ignore any air resistance and surface friction.

Answer

(i)   kinetic energy = ½ × mass × speed²   or $KE = \dfrac{1}{2} \times m \times v^2$

(ii)   **Step 1**   List all the information in symbol form and change into appropriate and consistent SI units if required.

$m = 45\,\text{kg}$
$v = 4.6\,\text{m/s}$
$KE = ?$

Step 2   Calculate the answer by putting the numbers into the equation.

$$KE = \dfrac{1}{2} \times 45 \times 4.6^2 = 476.1 = 480\,\text{J} \quad \text{(to 2 sig. figs)}$$

(iii)   gravitational potential energy = mass × gravitational field strength × height

or $GPE = m \times g \times h$

(iv)   **Step 1**   List all the information in symbol form and change into appropriate and consistent SI units if required.

GPE gained = KE lost = $480\,\text{J}$
$m = 45\,\text{kg}$
$g = 10\,\text{N/kg} = 10\,\text{m/s}^2$
$h = ?$

Step 2   Rearrange the equation.

$$GPE = m \times g \times h \quad \Rightarrow \quad h = \dfrac{GPE}{m \times g}$$

Step 3   Calculate the answer by putting the numbers into the equation.

$$h = \dfrac{480}{45 \times 10} = 1.066 = 1.1\,\text{m} \quad \text{(to 2 sig. figs)}$$

ALWAYS REMEMBER TO STATE THE UNIT FOR CALCULATED QUANTITIES.

When all the **KE** has been converted to **GPE**, the skateboarder has reached a height of 1.1 m. As the ramp is 1.3 m high, he does not reach the top.

- In any process the energy tends to **dissipate** (spread) into the surroundings. When a pendulum bob is pulled to one side, it is given gravitational potential energy because its height increases. When it is released, this begins to convert to kinetic energy, and then, as it swings past the centre of oscillation, it begins to convert back to potential energy. As time progresses the swings become smaller; some of the potential/kinetic energy is being transferred to the surroundings by heating because of air resistance. This energy is not recoverable.

- Energy changes are often multi-stage. Consider a hydroelectric power station (see the diagram at the bottom of page 150).
  - The water is stored in a dam at a higher level than the turbines of a generator; it has gravitational potential energy.
  - The water travels through tunnels or pipes to the turbines and its gravitational potential energy changes to kinetic energy as the water loses height.
  - The water causes the turbines to rotate; the kinetic energy of the water converts to kinetic energy of the turbines.
  - The rotation of the turbines is transferred to the generator and electricity is produced.
  - The energy is transferred electrically via power lines to the consumer; it can then be used in many ways.

  Remember: That the total energy remains the same although some energy will be dissipated or 'lost' to the surroundings in each transfer.

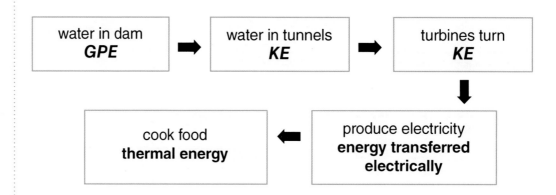

Power

- If a man pushes a weight through a distance, he does work. His **power** is related to how quickly he does that work; the faster he does it, the more power he has.

- A car does work when its engine exerts a driving force and it moves through a distance. Cars with more powerful engines can do work quicker and travel faster than less powerful ones.

- **Power** is the **rate of doing work**. It has the symbol **P**, and its unit is the **watt** (W); one watt is defined as **one joule per second**:

  1 W = 1 J/s

- **Remember:** work done = energy transferred
  Power is also the **rate of transfer of energy**. A 100 W lamp will transfer 100 J of energy every second.

- Power is related to energy transferred (work done) and the time taken for that transfer by the equation:

  $$\text{power} = \frac{\text{work done}}{\text{time taken}} = \frac{\text{energy transferred}}{\text{time taken}}$$

  $$P = \frac{W}{t} = \frac{E}{t}$$

  P = power (W)
  W = work done (J)
  t = time taken (s)
  E = energy transferred (J)

- It follows that **efficiency** (see page 125) can also be expressed by the following:

  $$\text{efficiency} = \frac{\text{useful power output}}{\text{total power input}} \times 100\%$$

Note

The **time taken** to do something **does not** have an effect on the **work done**. The time taken **does** have an effect on the **power**.

Example 4c (vi)

An athlete exerts an average horizontal force on the ground of 30 N while running. She runs a distance of 1.6 km in 5.0 minutes.

(i) State the equation linking work, force and distance.
(ii) Calculate the work the athlete does.
(iii) State the equation linking power, work done (or energy transferred) and time taken.
(iv) Calculate the power of the athlete.

Answer

(i) work done = force × distance moved   or $W = F \times d$

(ii) **Step 1** List all the information in symbol form and change into appropriate and consistent SI units if required.

$F = 30\,N$
$d = 1.6\,km = 1600\,m$
$W = ?$

Step 2 Calculate the answer by putting the numbers into the equation.

$W = 30 \times 1600 = 48000\,J$

(iii) $\text{power} = \dfrac{\text{work done}}{\text{time taken}} = \dfrac{\text{energy transferred}}{\text{time taken}}$ or $P = \dfrac{W}{t} = \dfrac{E}{t}$

(iv) **Step 1** List all the information in symbol form and change into appropriate and consistent SI units if required.

$W = 48000\,J$
$t = 5.0 \times 60\,s = 300\,s$
$P = ?$

Step 2 Calculate the answer by putting the numbers into the equation.

$P = \dfrac{48000}{300} = 160\,W$

ALWAYS REMEMBER TO STATE THE UNIT FOR CALCULATED QUANTITIES.

Section 4d Energy resources and electricity generation

- Energy cannot be created or destroyed; it can be **transferred** from one form into another.

- Electricity generation requires a **source** of energy.

- Almost all of our energy comes initially from the **Sun**. The exceptions are geothermal energy, energy from nuclear fission and tidal energy. The Sun's energy comes from nuclear fusion.

- **Fossil fuels** (**coal**, **oil** and **gas**) are currently the main source of energy used worldwide. The **chemical energy** in fossil fuels can be used to produce electricity.

- Fossil fuels are formed from the very highly compressed remains of dead plants and animals that lived many millions of years ago. Fossil fuel reserves will eventually run out. They are a **finite** or **non-renewable** resource.

- Energy sources are **renewable** or **non-renewable**:

Renewable	Non-renewable
waves	oil
solar	coal
tidal	gas
wind	uranium (nuclear fission)
hydroelectric	
geothermal	

Nuclear fission and nuclear fusion

- **Nuclear fission**, which occurs in nuclear power stations, is the process in which large nuclei like uranium **split** into smaller ones, releasing an enormous quantity of energy.

- **Nuclear fusion**, which occurs in the Sun, is the process in which two small nuclei **combine** to form a larger one, again releasing a huge amount of energy. When hydrogen atoms within stars combine to form helium the process is called fusion (see page 235).

Generation of electricity from fossil and nuclear fuels

○ **Chemical** energy from fuel is used to produce electricity. There are essentially four main steps to generating electricity in a power station when using fossil fuels (coal, oil or gas) or nuclear fuel. After generation, the electricity is transmitted all over the country (see page 215).

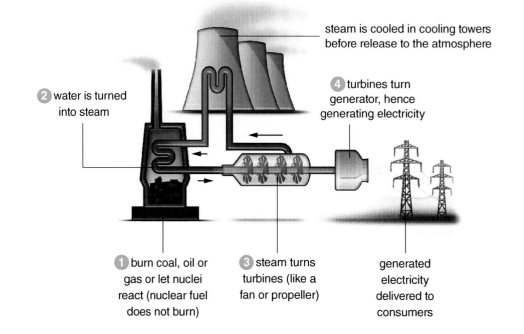

1. burn coal, oil or gas or let nuclei react (nuclear fuel does not burn)
2. water is turned into steam
3. steam turns turbines (like a fan or propeller)
4. turbines turn generator, hence generating electricity

steam is cooled in cooling towers before release to the atmosphere

generated electricity delivered to consumers

Generation of electricity from renewable sources

○ **Energy from the Sun**
Solar cells, also known as photovoltaic (PV) cells, transfer energy received by radiation (light) directly into electricity. Huge arrays of cells are needed for large-scale electricity generation.

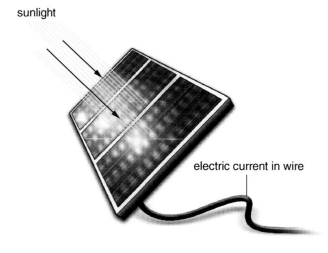

○ Small-scale arrays of photovoltaic cells are used in many different ways, from providing energy for a calculator to powering agricultural equipment in remote places or satellites orbiting the Earth.

○ **_Energy from wind_**
Kinetic energy of moving air (wind) is used to turn the rotor blades in a wind turbine. The turbines rotate the generator. The kinetic energy is converted into electricity by the generator. 'Farms' of many wind turbines are needed for large-scale electricity generation.

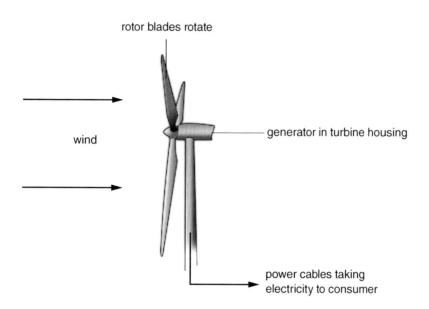

○ **_Energy from waves_**
Generators can be driven by the transverse (up and down) motion of ocean waves. The kinetic energy causes water to rise and fall in the air chamber (see diagram). The motion of the air above the water causes the turbine to turn and electricity is produced by the generator.

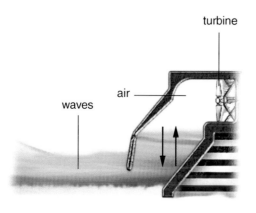

○ **Energy from tides**

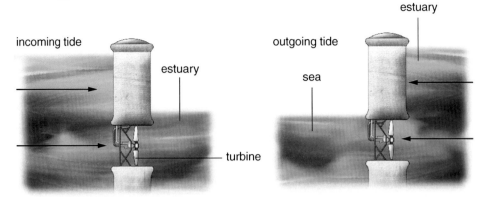

Tidal effects (the incoming and outgoing tides) cause the water to move. As it does so, it turns the turbines and electricity is generated. The gravitational potential energy and kinetic energy of the water drive the generators that produce electricity.

○ **Geothermal energy**

Water is fed down through pipes several kilometres underground, passing through hot rocks. These rocks heat the water until it turns into very hot steam. The steam is fed back up to ground level and is used to turn turbines and generate electricity.

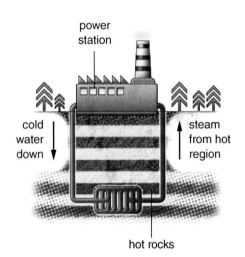

○ **Energy from trapped water**

A dam is built to trap water, sometimes in a valley where there is an existing lake. Water is allowed to flow through tunnels in the dam, to turn turbines and thus drive the generators that produce electricity. This is called **hydroelectricity**.

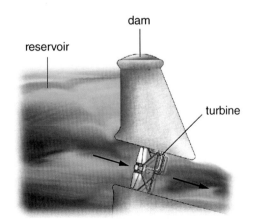

Solar heating

○ Energy transferred by radiation from the Sun can be used to heat water as shown in the diagram below. Solar water-heating panels can be fixed on roofs to capture this energy efficiently. The Sun heats the water in the solar panel and the hot water is pumped to the storage cylinder. Cold water takes its place and the cycle continues, with the water in the storage tank becoming ever hotter. There is usually an additional separate heating system with a more conventional boiler because, of course, solar panels only work during hours of daylight.

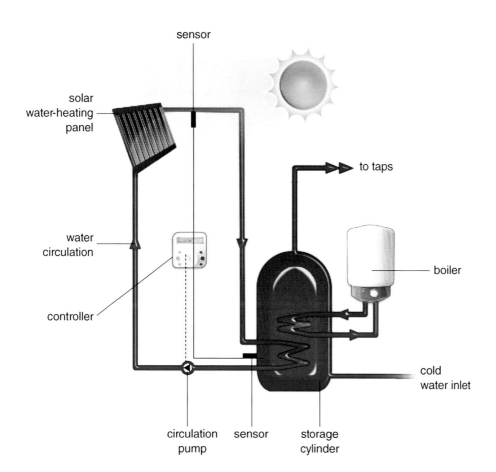

Comparing methods of generating electricity

Advantages of non-renewable sources

○ These fuels are far more 'energy dense' than renewable sources and thus, in many countries, they allow production of larger amounts of energy in comparison to renewable sources. They can be transported and stored ready for use.

Disadvantages of non-renewable sources

○ Fossil fuels are becoming increasingly expensive to mine (coal) and drill for (oil and gas), as reserves are running out (being **depleted**).

○ Fossil fuels cause **global warming** due to high carbon dioxide (CO_2) emissions when they are burnt. Some also produce sulfur dioxide (SO_2), which causes **acid rain**.

○ Nuclear power stations are expensive to build and any radiation leak or explosion may have a devastating and long-lasting effect on the local population and environment. The use of nuclear fuel produces radioactive waste products.

Advantages of renewable sources

○ They are all regarded as clean (producing little or no pollution).

○ They will not run out.

○ The fuel itself is cheap or free.

Disadvantages of renewable sources

○ Generally, renewable energy sources have high initial installation costs and the supply may not be constant.

○ *Energy from the Sun*
   - when the Sun doesn't shine (e.g. at night), no electricity is produced
   - dirty solar panels are inefficient
   - installing panels is an expensive process
   - the panels take up large areas.

○ *Energy from wind*
   - when the wind doesn't blow, no electricity is produced
   - wind farms can have an impact on the natural beauty of a landscape and may be considered noisy
   - offshore wind farms are expensive to build and need to be avoided by shipping.

- **Energy from waves**
    - waves vary in size and therefore will produce varying quantities (amounts) of energy; when the sea is calm, no electricity is produced
    - installation is expensive and challenging
    - people in charge of boats have to be aware of the location of the turbines to prevent accidents.

- **Energy from tides**
    - many countries do not have suitable locations
    - might affect local marine life and destroy habitats.

- **Geothermal energy**
    - there are few locations that are suitable, as the hot rocks need to be relatively close to the surface or in volcanic areas
    - it is often necessary to drill very deeply and this makes the energy very expensive to obtain.

- **Energy from trapped water (hydroelectricity)**
    - the local environment may be destroyed, as water needs to be stored behind a dam, and the land behind it is flooded
    - dams are expensive to build.

Section 5 Solids, liquids and gases

Section 5a Units

❑ In this section you will come across the following units:
- the **metre** (m) is the unit of length
- the **kilogram** (kg) is the unit of mass
- the **second** (s) is the unit of time
- the **metre squared** (m^2) is the unit of area
- the **metre cubed** (m^3) is the unit of volume
- the **kilogram per metre cubed** (kg/m^3) is the unit of density
- the **newton** (N) is the unit of force
- the **pascal** (Pa) is the unit of pressure
- the **joule** (J) is the unit of energy
- the **degree Celsius** (°C) is the unit of temperature
- the **kelvin** (K) is the unit of absolute temperature
- the **metre per second** (m/s) is the unit of speed or velocity
- the **metre per second squared** (m/s^2) is the unit of acceleration

○ And also:
- the **joule per kilogram degree Celsius** (J/(kg°C)) is the unit of specific heat capacity.

Section 5b Density and pressure

Density

❑ Density is defined as the **mass per unit volume** (how much mass is packed into a unit volume).

❑ Density is measured in kg/m^3. The symbol for density is **ρ**, which is the Greek letter rho.

❑ Mass has the symbol **m**, and its unit is the **kilogram** (kg).

❑ Volume has the symbol **V**, and is measured in **metres cubed** (m^3).

❑ These quantities are related by the formula:

$$\text{density} = \frac{\text{mass}}{\text{volume}}$$

$$\rho = \frac{m}{V}$$

ρ = density (kg/m^3)
m = mass (kg)
V = volume (m^3)

5b Density and pressure

> **Note**
>
> Each substance has its own density; one bar of gold will have the **same density** as 100 bars of gold. But the density will depend on the state of the substance (whether it is solid, liquid or gas).

- Generally, the density of a material when it is a solid is greater than its density when it is liquid.

- Water is an exception because the density of the solid, ice, is lower than that of water. Ice floats in water because it is less dense.

- Lead is more dense than water and sinks. Some wood is less dense than water and floats.

- A liquid of lower density will float on top of a liquid of higher density, e.g. oil floats on water.

- The density of a liquid is greater than that of a gas.

- Consider the following diagrams representing gas particles in a box. All the particles have the same mass.

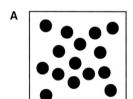 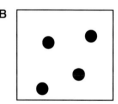

The gases in boxes A and B are contained in the same volume but have a different number of particles. The gas in box A has a higher density as it has more particles.

The gas in box C is half the volume of that in box D with the same number of particles; hence the gas in box C has the higher density.

The density of a gas depends on the number of particles (the mass) and the volume.

Determining density

❑ **Determining the density of a regularly shaped object**

Method

1. Use an electronic balance to measure the mass **m** of the object.

2. Use a ruler to measure the dimensions of the object and then calculate its volume **V**.

3. Then use the following equation to calculate density:

$$\rho = \frac{m}{V}$$

> **Note**
>
>
> If you are using a spring balance (newton meter) instead of an electronic balance, then you are measuring the weight **W**.
>
> Calculate the mass from the weight value using the formula:
>
> $$m = \frac{W}{g}$$ where **g** is the gravitational field strength (10 N/kg).

Example 5b (i)

The block of wood has a mass of 80 g.

(i) State the equation linking mass, volume and density.
(ii) Calculate the density of the wood.

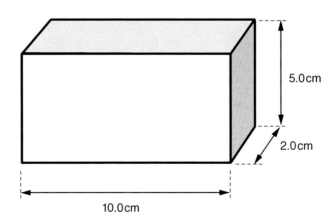

Answer

(i) density = $\frac{\text{mass}}{\text{volume}}$ or $\rho = \frac{m}{V}$

(ii) **Step 1** List all the information in symbol form and change into appropriate and consistent SI units if required.

$l = 10.0\,\text{cm}$
$w = 2.0\,\text{cm}$
$h = 5.0\,\text{cm}$
$m = 80\,\text{g}$
$\rho = ?$

Step 2 Use the correct equations.

$V = l \times w \times h$ and $\rho = \dfrac{m}{V}$

Step 3 Calculate the answer by putting the numbers into the equations.

$V = l \times w \times h = 10.0 \times 2.0 \times 5.0 = 100\,\text{cm}^3$

$\rho = \dfrac{m}{V} = \dfrac{80}{100} = 0.80\,\text{g/cm}^3$ (to 2 sig. figs)

ALWAYS REMEMBER TO STATE THE UNIT FOR CALCULATED QUANTITIES.

The unit g/cm³ is acceptable for density, although the SI unit is kg/m³. This is why cm has not been converted into m in the above calculation.

Determining the density of a liquid

❏ A measuring cylinder is used to measure the volume of a liquid. Volume must be measured to the bottom of the **meniscus** (the bottom of the curved surface of the liquid).

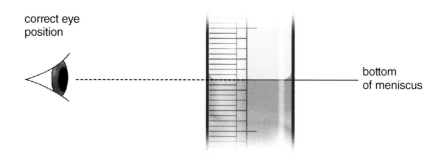

❏ When using a measuring cylinder, be careful to avoid parallax error (see pages 2-3). If the volume is not measured at right angles to the meniscus, parallax error will cause incorrect high or low measurements.

Method

1. The mass *m* of the liquid can be measured using an electronic balance. To find the mass of the liquid, we subtract the mass of the empty measuring cylinder from the mass of the liquid and the measuring cylinder.
2. The volume *V* can be read directly from the measuring cylinder.
3. Then use the following equation to calculate density.

$$\rho = \frac{m}{V}$$

Example 5b (ii)

(i) State the equation linking density, mass and volume.
(ii) Calculate the density of water using the information in the diagrams below.

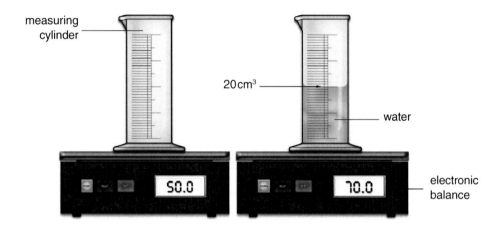

Answer

(i) $\text{density} = \frac{\text{mass}}{\text{volume}}$ or $\rho = \frac{m}{V}$

(ii) **Step 1** List all the information in symbol form and change into appropriate and consistent SI units if required.

$m = 70.0 - 50.0 = 20.0 \text{ g}$
$V = 20 \text{ cm}^3$
$\rho = ?$

Remember: It is acceptable to use mass in g and volume in cm^3 for density calculations.

Step 2 Calculate the answer by putting the numbers into the equation.

$$\rho = \frac{20.0}{20} = 1.0 \, g/cm^3$$

ALWAYS REMEMBER TO STATE THE UNIT FOR CALCULATED QUANTITIES.

❏ **Determining the density of an irregularly shaped object**

In the procedure below we use the **displacement** of water to determine the volume of an irregular object.

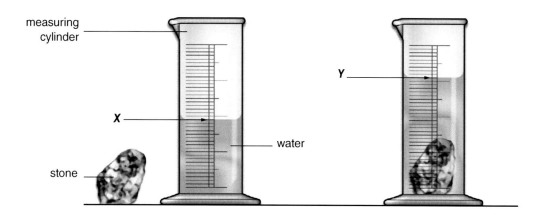

Method

1. Measure the mass *m* of the irregularly shaped object, in this case a stone, using an electronic balance.
2. Partially fill a measuring cylinder with a known volume **X** of water.
3. Lower the stone gently into the water until it is completely immersed, taking care not to lose water due to splashing.
4. Measure the new volume **Y**.
5. The volume of the stone is **V = Y − X**.
6. Use the following equation to calculate the density.

$$\text{density} = \frac{\text{mass}}{\text{volume}} \quad \text{or} \quad \rho = \frac{m}{V}$$

Remember: Always measure to the bottom of the meniscus when using a measuring cylinder.

Example 5b (iii)

An irregularly shaped piece of steel has a mass of 100 g. The steel is immersed in the water as shown.

(i) State the equation linking density, mass and volume.

(ii) Calculate the density of steel using the information from the diagrams below.

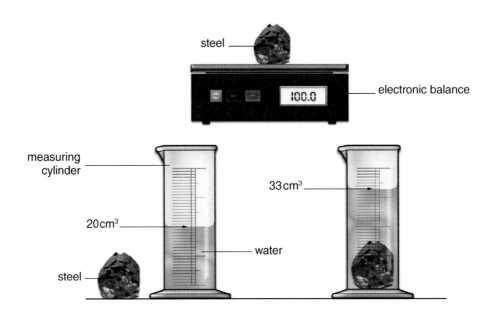

Answer

(i) density = $\dfrac{\text{mass}}{\text{volume}}$ or $\rho = \dfrac{m}{V}$

Step 1 List all the information in symbol form and change into appropriate and consistent SI units if required.

$m = 100\,\text{g}$
$V = 33 - 20 = 13\,\text{cm}^3$
$\rho = ?$

Step 2 Calculate the density by putting the numbers into the equation.

$\rho = \dfrac{100}{13} = 7.69 = 7.7\,\text{g/cm}^3$ (to 2 sig. figs)

ALWAYS REMEMBER TO STATE THE UNIT FOR CALCULATED QUANTITIES.

Pressure

- Pressure is defined as the **force per unit area**.

- Pressure has the symbol **p** and its unit is the **pascal** (Pa).

- Pressure, force and area are related by the equation:

    $$\text{pressure} = \frac{\text{force}}{\text{area}}$$

    $$p = \frac{F}{A}$$

    p = pressure (Pa) or (N/m^2)
    F = force (N)
    A = area (m^2)

- 1 Pa is equivalent to 1 N/m^2 (**newton per metre squared**).

- Pressure can be increased by **increasing the force** on a constant area.

    Pressure can be increased by **decreasing the area** for a constant force.

- A girl weighing 500 N and wearing high heels with a combined area of 2 cm^2 in contact with the floor can make indentations on a wooden floor, because the pressure she exerts is 2 500 000 Pa (2.5 MPa).

    An elephant weighing 40 000 N and standing on all four feet, a total area of 0.4 m^2, would exert a pressure of only 100 000 Pa.

- A sharp knife cuts bread more easily than a blunt knife. This is because the sharp knife has a much smaller surface area in contact with the bread than a blunt knife. When you push down on the knife (exert a force), the sharp knife exerts greater pressure on the bread and it cuts easily.

Example 5b (iv)

A rectangular block has a mass of 220 g and dimensions of 12.0 cm × 5.0 cm × 2.0 cm.

(i) State the equation linking pressure, force and area.
(ii) State the equation linking weight, mass and gravitational field strength.
(iii) Calculate the maximum pressure that can be exerted by the block on the surface on which it rests.

Take the gravitational field strength to be 10 N/kg.

5 Solids, liquids and gases

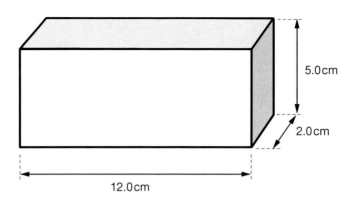

Answer

(i) pressure = $\dfrac{\text{force}}{\text{area}}$ or $p = \dfrac{F}{A}$

(ii) weight = mass × gravitational field strength or $W = m \times g$

(iii) **Step 1** List all the information in symbol form and change into appropriate and consistent SI units if required.

The question asks for the maximum pressure. The **maximum** pressure is exerted when the block rests on the **smallest** area.

$m = 220\,\text{g} = 0.22\,\text{kg}$
$A = 5.0 \times 2.0 = 10\,\text{cm}^2 = 10 \times 10^{-4}\,\text{m}^2 = 1.0 \times 10^{-3}\,\text{m}^2$
$g = 10\,\text{N/kg}$
$p = ?$

Remember: Take care when changing the units of area:

$1\,\text{cm}^2 = \dfrac{1}{100} \times \dfrac{1}{100} = 1 \times 10^{-4}\,\text{m}^2$

Step 2 Calculate the answer by putting the numbers into the equations, remembering that F = weight.

$W = m \times g = 0.22 \times 10 = 2.2\,\text{N}$

$p = \dfrac{F}{A} = \dfrac{2.2}{1.0 \times 1.0^{-3}} = 2200\,\text{Pa}$

ALWAYS REMEMBER TO STATE THE UNIT FOR CALCULATED QUANTITIES.

More practical examples of pressure

- Snow shoes shaped like tennis rackets are used in snow-covered places so that people don't sink into the snow. The **large area** of the snow shoe **decreases** the **pressure** on the snow.

snow shoes

- Camels have **large feet** to **decrease** the **pressure** they exert on the ground so that they do not sink into the desert sand.

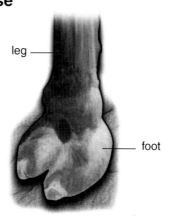
leg
foot

- Pushing a drawing pin into soft wood is easy as the point has a **very small area** so there is a **very large pressure**.

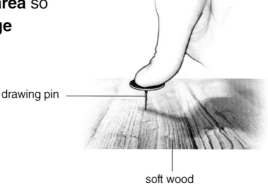
drawing pin
soft wood

Atmospheric pressure

❑ The air around us exerts a pressure. This pressure is called **atmospheric pressure**.

❑ Atmospheric pressure has the following properties.
- Its effect acts **equally in all directions**.
- It **decreases** as altitude (height above sea level) **increases** because the air molecules become further apart (the air is less dense), and vice versa.
- At **sea level**, atmospheric pressure is approximately 100 000 Pa (100 kPa).

Liquid pressure

❑ Liquids also exert pressure. Pressure in liquids has the following properties.
- The pressure acts **equally in all directions**; it causes a force to push on every surface that the liquid is in contact with.
- The pressure increases with **increasing depth** due to the weight of the column of liquid (height h) above the measurement point.
- The pressure also depends on the **density** of the liquid; the greater the density, the greater the pressure at a given depth.

❑ The pressure of a liquid varies directly with the height h and the density ρ (rho) of the liquid. As a result the pressure will be greater at the bottom of a 10 cm tall measuring cylinder full of water than the pressure due to 10 cm of oil in an identical cylinder because the water is more **dense**.

To achieve the same pressure with oil, you would need a taller cylinder to increase the height h.

❑ In a liquid, pressure difference, density, gravitational field strength and height are related by the equation:

$$\frac{\text{pressure}}{\text{difference}} = \text{height} \times \text{density} \times \frac{\text{gravitational}}{\text{field strength}}$$

$$p = h \times \rho \times g$$

p = pressure difference (Pa)
h = height (m)
ρ = density (kg/m³)
g = gravitational field strength (N/kg)

Remember:

density = $\frac{\text{mass}}{\text{volume}}$ or $\rho = \frac{m}{V}$

❑ By rearranging the equation for pressure we can calculate the height (or depth) of liquid:

$h = \frac{p}{\rho \times g}$

Example 5b (v)
A diver is 18m below the surface of water of density 1000kg/m³.
(i) State the equation linking pressure difference to the density of the liquid, the height/depth of the liquid and the gravitational field strength.
(ii) Calculate the pressure that the water exerts on the diver.

Answer
(i) $\frac{\text{pressure}}{\text{difference}}$ = height × density × $\frac{\text{gravitational}}{\text{field strength}}$ or $p = h \times \rho \times g$

(ii) **Step 1** List all the information in symbol form and change into appropriate and consistent SI units if required.

$h = 18\,\text{m}$
$\rho = 1000\,\text{kg/m}^3$
$g = 10\,\text{N/kg}$
$p = ?$

Step 2 Calculate the answer by putting the numbers into the equation.

$p = 18 \times 1000 \times 10 = 180000\,\text{Pa} = 1.8 \times 10^5\,\text{Pa}$

ALWAYS REMEMBER TO STATE THE UNIT FOR CALCULATED QUANTITIES.

Top Tip

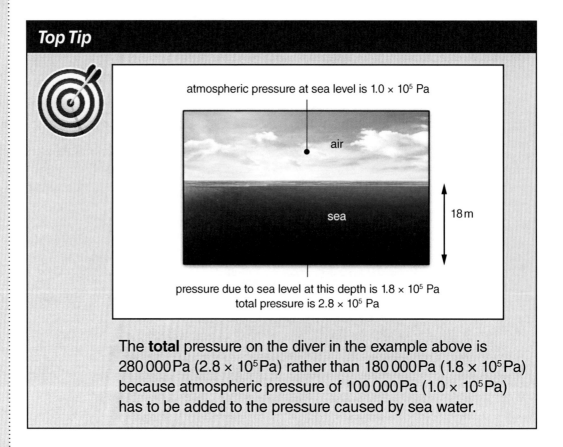

The **total** pressure on the diver in the example above is 280 000 Pa (2.8×10^5 Pa) rather than 180 000 Pa (1.8×10^5 Pa) because atmospheric pressure of 100 000 Pa (1.0×10^5 Pa) has to be added to the pressure caused by sea water.

Section 5c Change of state

○ The three **states of matter** are:
1. solid
2. liquid
3. gas.

○ The kinetic model of matter explains the behaviour of solids, liquids and gases in terms of how the **particles** (molecules or atoms) from which they are made are arranged, and how they move.

Properties of a solid

○ In a solid:
- The particles are arranged **close together** in a regular lattice pattern, for example arranged in neat rows.
- The particles **vibrate** around a fixed position but do not move from place to place.

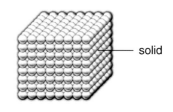

- Solids have a **fixed shape** and a **fixed volume** (provided the temperature and pressure remain constant).

- There are very **strong forces of attraction** between particles because they are close together. (Individual particles cannot break free from the lattice.)

- The solid cannot be squashed or compressed easily because the particles are close together and it takes a very large force to push them closer.

Properties of a liquid

- In a liquid:
  - The particles are close together but have no fixed arrangement.
  - The particles are **free to move** and tend to **slip and slide past each other**.

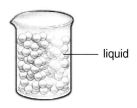

- Liquids do not have a fixed shape; they **take the shape** of the bottom of the container they are in. They have a level and horizontal surface.

- Liquids have a **fixed volume** (provided the temperature and pressure remain constant).

- There are slightly **weaker forces of attraction** between particles in a liquid than in a solid.

- Liquids cannot be squashed or compressed because the particles are close together.

Properties of a gas

- In a gas:
  - The particles are very much further apart than in liquids.
  - The particles **move very fast** in **random directions**.
  - The particles are constantly **colliding** with each other and the walls of the container.

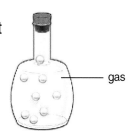

5 Solids, liquids and gases

- Gases have **no fixed shape** or **volume**; they fill any container in which they are placed.
- Gases can be squashed or **compressed** because the particles are far apart.
- There are **negligible** forces of attraction between particles.

> **Note**
>
> The **temperature of a substance** is a measure of the **kinetic energy** of its particles. The **faster** the particles move or vibrate, the greater the temperature and the **hotter** the substance. The **slower** the particles move or vibrate, the **colder** the substance.

The table below lists the main properties of solids, liquids and gases.

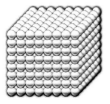  

Solid	Liquid	Gas
very slightly compressible	incompressible	can be compressed
fixed shape	takes on the shape of the bottom of the container	fills the container, taking on its shape
fixed volume	fixed volume	no fixed volume
static	can flow	can flow
particles very closely packed	particles disordered and closely packed	particles far apart
very strong forces of attraction between atoms/molecules	fairly strong forces of attraction between atoms/molecules	almost no forces of attraction between atoms/molecules
atoms/molecules vibrate about a fixed position	atoms/molecules slide past each other	atoms/molecules show fast and random motion

Note

Particles in the solid state have the least energy and particles in the gaseous state have the most energy.

Remember: When it comes to states of matter:

think movement – think energy.

Thermal energy

- **Thermal energy** is one of the forms of energy store (see page 123) and its unit is the **joule** (J). When substances **absorb** thermal energy, their temperature usually rises.

- **Specific heat capacity** (s.h.c.) is the amount of thermal energy needed to raise the temperature of **1 kg** of a substance by **1 °C**. It has the symbol **c** and its unit is the **joule per kilogram degree Celsius**, J/(kg °C).

- The amount of thermal energy required to raise the temperature of a substance without changing state (see pages 175-78) is:

$$\text{change in thermal energy} = \text{mass} \times \text{specific heat capacity} \times \text{change in temperature}$$

$$\Delta Q = m \times c \times \Delta T$$

ΔQ = change in thermal energy (J)
m = mass (kg)
c = specific heat capacity (J/(kg °C))
ΔT = change in temperature (°C)

Top Tip

On a hot summer's day in Spain the sand on a beach is very hot to step on; it is cold to step on at night. Sand has a **low specific heat capacity** and so the temperature change is large for a particular transfer of energy.

5 Solids, liquids and gases

Example 5c (i)
The specific heat capacity of water is 4200 J/(kg °C).
(i) State the equation linking change in thermal energy, mass, specific heat capacity and change in temperature.
(ii) Calculate the temperature change when 150 kJ of energy is given to 3.0 kg of water.

Answer
(i) change in thermal energy = mass × specific heat capacity × change in temperature

or $\Delta Q = m \times c \times \Delta T$

(ii) **Step 1** List all the information in symbol form and change into appropriate and consistent SI units if required.

$\Delta Q = 150\,\text{kJ} = 150\,000\,\text{J}$
$c = 4200\,\text{J/(kg°C)}$
$m = 3.0\,\text{kg}$
$\Delta T = ?$

Step 2 Rearrange the equation.

$\Delta Q = m \times c \times \Delta T \quad \Rightarrow \quad \Delta T = \dfrac{\Delta Q}{m \times c}$

Step 3 Calculate the answer by putting the numbers into the equation.

$\Delta T = \dfrac{150\,000}{3.0 \times 4200} = 11.90 = 12\,°C$ (to 2 sig. figs)

ALWAYS REMEMBER TO STATE THE UNIT FOR CALCULATED QUANTITIES.

Note

If 1 kg of lead is given the same amount of thermal energy as 1 kg of copper, then the temperature rise of the lead is about **three times greater** than that of copper. Lead's specific heat capacity is about **three times smaller** than copper's.

Determining specific heat capacity

○ **Remember:** Specific heat capacity is defined as the thermal energy needed to raise the temperature of 1 kg of a substance by 1°C.

○ The following experiment describes how to determine the specific heat capacity of a metal.

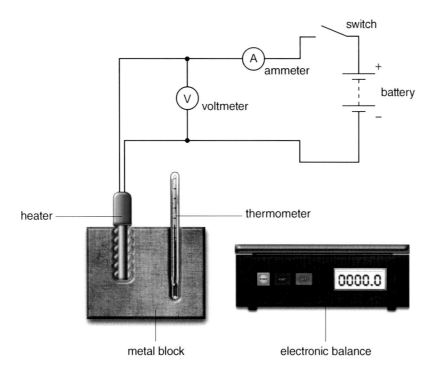

Method

1. Set up the experiment as shown.
2. Place the metal block on the electronic balance; record its mass **m**.
3. Record the initial temperature $T_{initial}$ of the block.
4. Switch the apparatus on, start a stopwatch and record the current and voltage readings.
5. After a time **t**, switch off the power supply and record the final temperature T_{final} of the block.

   Remember: If the time is in minutes, for example 6 minutes 30 seconds, it must be converted to seconds, i.e. 390 seconds.

Note

The specific heat capacity of water can be found in a similar way if the metal block is replaced by a cup of water. To find the mass of water using a balance, you would subtract the mass of the empty cup from the mass of the cup containing water. A lid should be added to the cup to avoid evaporation, and the liquid should be gently stirred so that the thermal energy is dispersed evenly throughout the liquid.

Calculations to be made

1. Energy transferred by the electric heater:

$$E = I \times V \times t$$

2. Change in temperature:

$$\Delta T = T_{final} - T_{initial}$$

3. Specific heat capacity of the metal block:

$$\Delta Q = m \times c \times \Delta T \implies c = \frac{\Delta Q}{m \times \Delta T}$$

4. From conservation of energy, $\Delta Q = E$, so

$$c = \frac{I \times V \times t}{m \times \Delta T}$$

Improvements to the experiment

- You may be asked to suggest improvements for practical experiments. In this case we are trying to find how thermal energy affects the block, therefore we want to make sure **minimal thermal energy is dissipated to the surroundings**.

  The experiment could have been improved by:

  1. adding lagging (insulation) on the sides, top and underneath the block
  2. repeating the experiment to take an average.

Assumptions in the experiment and sources of error

- We assume that all the energy supplied is used to heat the metal block. Some, however, is **dissipated** to the surroundings. The greater the temperature difference between an object and the surroundings, the greater the rate of dissipation of energy.

○ If effective insulation is used, however, the thermal energy dissipated to the surroundings is very small and can often be neglected.

> **Top Tip**
>
> Very often the specific heat capacity value obtained in calculations using experimental results is higher than the actual value; this is because energy is **dissipated to the surroundings**.

Calculating energy dissipated to the surroundings

○ The example below shows how to calculate the thermal energy dissipated to the surroundings, when this is not negligible.

Remember: Energy is conserved:
energy into system = energy out of system

Example 5c (ii)

A 1.0 kW immersion heater takes 15 minutes to raise the temperature of 1.0 kg of water by 26 °C.

(i) State the equation linking energy transferred electrically to power and time.
(ii) State the equation linking change in thermal energy to mass, specific heat capacity and change in temperature.
(iii) Calculate the change in thermal energy if the specific heat capacity of water is 4200 J/(kg °C).
(iv) Calculate the energy dissipated to the surroundings, giving your answer to 2 sig. figs.

Answer

(i) energy transferred = power × time taken or $E = P \times t$

(ii) change in thermal energy = mass × specific heat capacity × change in temperature

or $\Delta Q = m \times c \times \Delta T$

5 Solids, liquids and gases

(iii) **Step 1** List all the information in symbol form and change into appropriate and consistent SI units if required.

$$m = 1.0 \,\text{kg}$$
$$c = 4200 \,\text{J/(kg°C)}$$
$$\Delta T = 26°\text{C}$$
$$\Delta Q = ?$$

Step 2 Calculate the answer by putting the numbers into the equation.

Change in thermal energy of water:

$$\Delta Q = m \times c \times \Delta T = 1.0 \times 4200 \times 26 = 109\,200 \,\text{J}$$

(iv) **Step 1** List all the information in symbol form and change into appropriate and consistent SI units if required.

$$P = 1 \,\text{kW} = 1000 \,\text{W}$$
$$t = 15 \text{ minutes} = 15 \times 60 = 900 \,\text{s}$$
$$\Delta Q = 109\,200 \,\text{J}$$
$$E = ?$$

Step 2 Calculate the answer by putting the numbers into the equation.

Energy transferred from the heater:

$$E = P \times t = 1000 \times 900 = 900\,000 \,\text{J}$$

Change in thermal energy of water:

$$\Delta Q = 109\,200 \,\text{J}$$

Energy dissipated $= E - \Delta Q = 900\,000 - 109\,200 \,\text{J}$

Energy dissipated $= 790\,800 \,\text{J} = 790 \,\text{kJ}$ (to 2 sig. figs)

ALWAYS REMEMBER TO STATE THE UNIT FOR CALCULATED QUANTITIES.

Change of state

Melting and boiling

○ Sometimes when substances absorb or lose thermal energy they change state.

○ The **changes of state** that can take place are:

boiling – a liquid changing to a gas
condensation – a gas changing to a liquid
solidification – a liquid changing to a solid
melting – a solid changing to a liquid

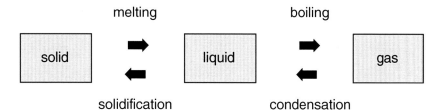

○ Energy must be **provided** for melting or boiling. Energy is **given out** during solidification or condensation.

○ The temperature of a substance **remains constant** while the substance changes state at its melting point or boiling point. The temperature remains the same until the change of state is complete.

○ The **melting point** is the temperature at which a solid turns into a liquid. The melting point of pure ice is 0 °C.

○ The **boiling point** is the temperature at which a liquid turns into a gas. The boiling point of pure water is 100 °C.

○ When melting or boiling, all the supplied energy is being used to **weaken** or **break** the bonds (forces of attraction) between molecules.

○ When condensation occurs, the molecules slow down and the bonds are **strengthened** or **formed**. These bonds bring the molecules closer together and the substance becomes liquid.

5 Solids, liquids and gases

○ When solidification occurs, the temperature drops and fewer molecules have enough kinetic energy to overcome neighbouring attractions. The molecules can only vibrate, the substance gains a shape of its own and becomes a solid.

○ The amount of energy needed to change the state of a substance depends only on the **mass** and the **type of substance**. Melting a substance requires a different amount of heat from that required to boil it.

○ The energy required to change a substance from a **solid** to a **liquid** at its melting point (without change of temperature) is the **latent heat of fusion**. (Fusion is an old word for melting.) At the melting point the energy of the molecules increases and they move further apart until the forces of attraction are reduced and the solid becomes liquid.

○ The energy required to change a substance from a **liquid** to a gas at its boiling point (without change of temperature) is the **latent heat of vaporisation**. At the boiling point the molecules move even further apart until there are no forces of attraction between them.

○ The latent heat of vaporisation of a material is greater than the latent heat of fusion because it requires more energy to completely separate the molecules (liquid to gas) than to reduce the forces of attraction (solid to liquid).

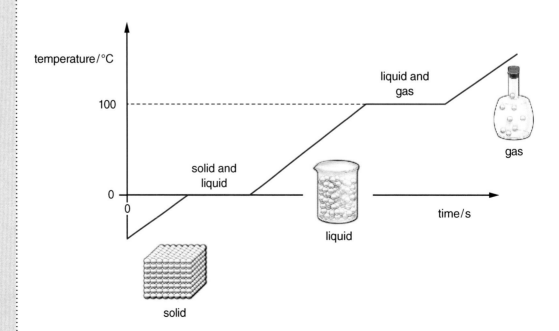

- The previous graph shows how the temperature changes with time when ice is heated.
  - While the ice is solid, its temperature increases as thermal energy is supplied.
  - When the ice reaches 0°C the temperature remains constant as all the supplied energy is used to change the state from solid (ice) to liquid (water).
  - The water changes temperature from 0°C to 100°C as thermal energy continues to be supplied.
  - At 100°C the temperature remains constant as the water is turned to steam.
- When a substance cools down and changes state from gas to liquid, or from liquid to solid, the latent heat is given out to the surroundings.

Top Tip

The **temperature does not increase** while a substance is **changing state** from solid to liquid or from liquid to gas because all the energy is being used to overcome the **forces of attraction** between molecules.

Boiling and evaporation

- Boiling and evaporation both describe a change of state from liquid to gas or vapour. (A vapour is a substance in its gaseous state that can be liquefied by pressure alone i.e. without cooling.)
- When a liquid boils, **all** the liquid molecules convert to gas molecules.
- Evaporation is the escape of the **most energetic molecules** from the **surface** of a liquid. Not all molecules in a liquid have the same energy. It is those molecules that are moving the fastest (i.e. those with the greatest kinetic energy) and are at the surface that can escape.

- If the more energetic molecules leave the liquid, it follows that the average energy of the molecules of the liquid falls and so its temperature falls. Evaporation **leads to the cooling** of a liquid.

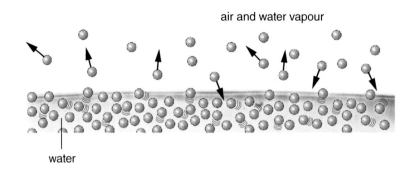

- Because of the random motion, some vapour molecules return to the surface of the liquid unless they are removed by draught or wind.

- Evaporation is the reason why wet clothes dry on a washing line, or a saucer of water eventually dries up.

- Boiling and evaporation differ in the following ways.
  - Boiling occurs **throughout** the liquid, whereas evaporation only occurs at the **surface** of the liquid.
  - Boiling only occurs at **one temperature** (100°C for water at standard atmospheric pressure), whereas evaporation occurs at **all temperatures**.
  - Boiling requires a supply of energy for the liquid to reach the required temperature, whereas evaporation occurs at all temperatures, utilising the energy stored in the liquid.
  - Boiling is usually a rapid process. Evaporation is usually a slow process but can be speeded up by a draught over the surface of the liquid or by an increased surface area in contact with the air.

Section 5d Ideal gas molecules

Pressure caused by gases

- Gas molecules move randomly and have large **kinetic energy** (because they are moving very fast). The molecules do not all have the same energy; some have more than others.
  The **average kinetic energy** is related to the **temperature** of the gas. We will see later (see page 184) that the average kinetic energy is directly proportional to the absolute temperature.

- **Increasing the temperature** of a gas increases the **average kinetic energy** of molecules within that gas and they move **faster**.

- The gas molecules exert a **force** on the walls of a container when they collide with it. The **pressure** exerted by the gas molecules is the **force per unit area**. The total pressure of the gas is the effect of the sum of all the collisions with the walls.

- When the molecules collide with the walls of the container, they change direction and bounce back. This means that the velocity (a vector quantity) has changed. If the velocity changes, there is an acceleration, and as $F = m \times a$ there is a force (see pages 37-38 and 184).

- In an **ideal gas**, it is assumed that there is no interaction between the molecules, and that when the molecules collide with the walls of the container they do not lose kinetic energy. In reality, gases are not ideal but they do approximate to this behaviour.

- The force is **larger** if the molecules are moving **faster** or if there are **more molecules** colliding with the walls **per second**.

- A **higher temperature** causes **more** collisions in a given time and causes the collisions to be **harder** because the molecules are moving faster. Consequently, a higher temperature causes a **larger pressure**.

Top Tip

As the **temperature** of a gas **increases**, the **pressure increases** because:
- the molecules have **greater kinetic energy**
- they hit the walls of the container **harder**
- they hit the walls of the container **more often**.

Temperature

❑ Temperature is the measure of how **hot** an object is. The hotness of an object is a measure of the **average kinetic energy** of the molecules that make up that object. When a substance is heated (without a change of state), the average speed (and the kinetic energy, which depends on square of speed) of the molecules increases. When a substance is cooled, the average speed of the molecules decreases.

❑ Theoretically, a temperature can be reached at which all motion of particles has ceased. This temperature is called **absolute zero**, which is $-273.15\,°C$ (usually taken as $-273\,°C$.)

❑ In the mid-1800s, a scientist named Lord Kelvin developed a scale of temperature known as the **Kelvin scale** or the **absolute** scale and he took absolute zero as the starting point.

❑ Temperature on the absolute scale has the symbol T, and its unit is the **kelvin** (K). Absolute zero is 0 K.

❑ The size of one degree Celsius is the same as one kelvin, so it follows that if:

$$0\,K = -273\,°C$$
$$273\,K = 0\,°C$$

and then $\quad 283\,K = 10\,°C$ and so on.

Top Tip

To convert from Celsius to kelvin:
temperature in $°C + 273$ = temperature in K

To convert from kelvin to Celsius:
temperature in $K - 273$ = temperature in $°C$

Pressure changes in a gas

❑ Important variables for a gas are:
1. pressure
2. volume
3. temperature
4. mass.

Boyle's Law

❑ Boyle's Law explains how, for an ideal gas, the pressure and volume are related when the **temperature and mass of gas are kept constant**.

❑ If the piston of a bicycle pump with a sealed end is pushed in at constant temperature, then the further you push it in, the harder it gets to push.

❑ This is because the **pressure** inside the container (pump) **increases as the volume decreases**.

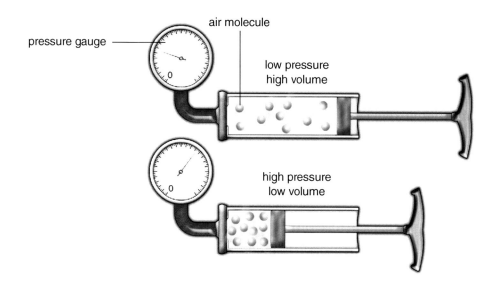

❑ The pressure increases because the gas molecules are squeezed into a smaller space. Therefore there are more collisions per second between the molecules and the container walls. The more frequent collisions cause a larger force on the walls and hence a larger pressure. The collisions do not get harder because the temperature does not change; it is the frequency of collisions that is responsible for the change in pressure.

5 Solids, liquids and gases

> **Top Tip**
>
> **When the volume of a fixed mass of gas is decreased at a constant temperature, the pressure increases.**
> This is because the molecules or particles are squeezed into a smaller space and therefore **collide more often** with the walls of the container and so the pressure increases. They **do not** move faster.

❏ **Boyle's Law** states that:

  For a fixed mass of an ideal gas at constant temperature, the pressure is inversely proportional to the volume.

  $$p \propto \frac{1}{V}$$

❏ This means that for a fixed mass of gas at constant temperature, **the pressure multiplied by the volume is constant.**

  $$p \times V = \text{constant}$$

❏ In other words, when **the pressure increases, the volume decreases** and vice versa, as shown in the graph below. Notice that the line does not touch either axis.

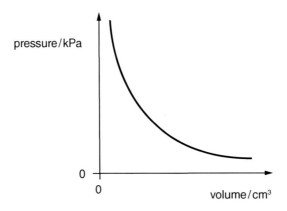

❏ In a change, initial and final values of pressure and volume are related by the equation:

  initial pressure × initial volume = final pressure × final volume

  $$p_1 \times V_1 = p_2 \times V_2$$

  p_1 = initial pressure (Pa or N/m²)
  V_1 = initial volume (cm³ or m³)
  p_2 = final pressure (Pa or N/m²)
  V_2 = final volume (cm³ or m³)

5d   Ideal gas molecules   Notes

Example 5d (i)

A bicycle pump contains 300 cm³ of air at atmospheric pressure. The air is compressed slowly.

(i) State the equation linking pressure and volume at constant temperature.
(ii) Calculate the pressure when the volume of the air is compressed to 125 cm³.
Atmospheric pressure is 100 kPa.

Answer

(i) initial pressure × initial volume = final pressure × final volume
or $p_1 \times V_1 = p_2 \times V_2$

Step 1   List all the information in symbol form and change into appropriate and consistent SI units if required.

$p_1 = 100\,\text{kPa} = 100\,000\,\text{Pa}$
$V_1 = 300\,\text{cm}^3$
$p_2 = ?$
$V_2 = 125\,\text{cm}^3$

Step 2   Rearrange the equation.

$$p_1 \times V_1 = p_2 \times V_2 \quad \Rightarrow \quad p_2 = \frac{p_1 \times V_1}{V_2}$$

Step 3   Calculate the answer by putting the numbers into the equation.

$$p_2 = \frac{100\,000 \times 300}{125} = 240\,000\,\text{Pa}$$

ALWAYS REMEMBER TO STATE THE UNIT FOR CALCULATED QUANTITIES.

N.B. Because the air is compressed slowly we can assume the temperature remains constant and apply Boyle's Law.

Increasing the temperature of a fixed volume gas

- The pressure of a **fixed volume** of gas is related to its temperature. As the temperature increases, so does the pressure.

- When the temperature of a gas is increased, the **kinetic energy** of its molecules **increases**. The molecules move on average faster because the average kinetic energy increases.

- For an ideal gas, the **average kinetic energy is directly proportional to the Kelvin (absolute) temperature**.

- The pressure of the gas is determined by the force of the molecules colliding with the walls of the container and by the rate of collision.

- As the temperature of the gas increases:
  - the force of each collision increases
  - the number of collisions every second increases, so the **pressure increases**.

- The relationship is referred to as the **Pressure Law**, and it states that:

For a fixed mass of ideal gas at constant volume, the pressure is directly proportional to the temperature measured in kelvin.

$p \propto T$

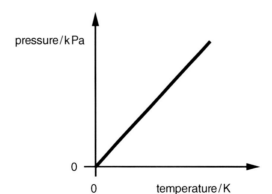

❏ In a change, initial and final values of pressure and temperature are related by the equation:

$$\frac{\text{initial pressure}}{\text{initial temperature}} = \frac{\text{final pressure}}{\text{final temperature}}$$

$$\boxed{\frac{p_1}{T_1} = \frac{p_2}{T_2}}$$

p_1 = initial pressure (Pa)
T_1 = initial temperature (K)
p_2 = final pressure (Pa)
T_2 = final temperature (K)

Top Tip

When the temperature of a fixed mass of gas is increased, its pressure increases if the volume is kept constant. This is because the average kinetic energy and therefore average velocity of the molecules increases and so the total force exerted on the walls of the container is greater.

Example 5d (ii)

A car tyre contains air at a pressure of 220 kPa and at a temperature of 20°C.

(i) State the equation linking the pressure and temperature of an ideal gas.

(ii) Calculate the pressure after the car has travelled a long distance when the temperature of the air in the tyre has increased to 35°C.

Answer

(i) pressure of a gas is directly proportional to its temperature in kelvin or $p \propto T$

$$\frac{\text{initial pressure}}{\text{initial temperature}} = \frac{\text{final pressure}}{\text{final temperature}} \quad \text{or} \quad \frac{p_1}{T_1} = \frac{p_2}{T_2}$$

(ii) **Step 1** List all the information in symbol form and change into appropriate and consistent SI units if required.

$p_1 = 220 \text{ kPa} = 220 \times 10^3 \text{ Pa}$
$T_1 = 20°C = 20 + 273 = 293 \text{ K}$
$p_2 = ?$
$T_2 = 35°C = 35 + 273 = 308 \text{ K}$

Step 2 Rearrange the equation.

$$\frac{p_1}{T_1} = \frac{p_2}{T_2} \implies p_2 = \frac{p_1 \times T_2}{T_1}$$

Step 3 Calculate the answer by putting the numbers into the equation.

$$p_2 = \frac{220 \times 10^3 \times 308}{293} = 231 \times 10^3 = 230\,\text{kPa}$$
(to 2 sig. figs)

ALWAYS REMEMBER TO STATE THE UNIT FOR CALCULATED QUANTITIES.

Increasing the temperature of a gas at constant pressure

- The volume of a **fixed mass** of gas at constant pressure is related to its temperature. As the temperature increases, so does its volume.

- When the temperature of a gas is increased, the **kinetic energy of its molecules increases**. The molecules on average move faster.

- The molecules collide with greater force on the walls of the container and more often but, because the walls of the container can expand, they do so until the gas pressure inside the container equals the external pressure. Thus the gas and the volume of the container expands as the temperature rises.

- A simple experiment can be done to show the relationship between the volume of a gas at constant pressure and its temperature.

In the diagram opposite, the narrow glass tube has a bubble of air trapped by liquid. The tube is open to the atmosphere at the top so the pressure on the gas is caused by atmospheric pressure and the liquid pressure. It remains constant throughout the experiment.

5d Ideal gas molecules

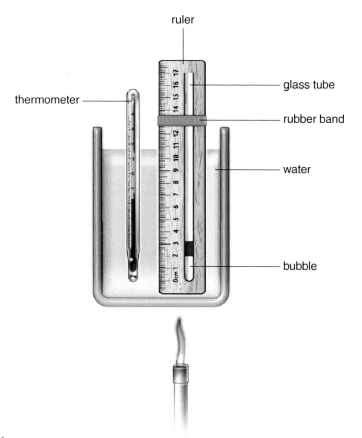

Method

1. Attach the glass tube to a ruler as shown and place in a water bath with a thermometer.

2. Measure and record the length of the bubble for a range of temperatures.
   N.B. The glass tube has a uniform cross-section, the length of the air bubble is proportional to the volume.

3. Plot a graph of length in cm against temperature in °C. Draw a best fit straight line and extrapolate the line back till it cuts the temperature axis.

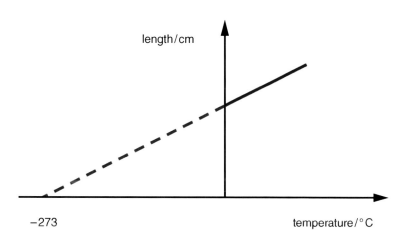

- The graph is a straight line, showing that, as the temperature increases, so does the length of the bubble (and therefore so does the volume). It does not pass through the origin, but it cuts the temperature axis at −273°C, absolute zero.

- If the temperature had been measured in kelvin, the graph would have been a straight line passing through the origin, showing that the **volume of the gas is directly proportional to the temperature in kelvin**.

    N.B. It doesn't matter what gas is used (provided it approximates to ideal), or what the original volume/length is. All graphs when extrapolated would go to −273°C theoretically.

Everyday examples

- When you use a pump to inflate a balloon, you add air to the balloon. The pressure inside the balloon increases because there are more molecules striking the walls of the balloon.

- A balloon that is blown up and knotted can be used to demonstrate the relationship between pressure and volume, and between volume and temperature.

- If you squeeze the balloon to reduce its volume, you can feel the increase in pressure on the walls of the balloon. This is because the air molecules strike the walls inside the balloon more often.

- If the balloon is placed on a hot radiator, the air inside will heat up and the balloon will expand. The pressure remains at atmospheric pressure but the volume increases as the temperature increases. This is because the air molecules are moving faster (they have more kinetic energy on average) and strike the wall of the balloon more often, pushing against the fabric of the balloon and making it expand because there is a pressure difference. The balloon will keep expanding until the pressure inside again equals the pressure outside.

- Conversely if the balloon is placed inside a freezer, the balloon will shrink as the temperature of the air decreases.

How a fire extinguisher works

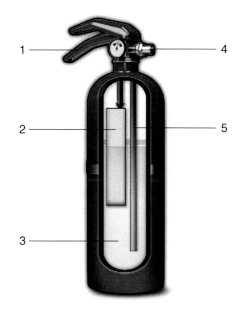

❑ The extinguisher in the diagram above contains water (3) and carbon dioxide as a compressed gas (2) in a sealed tube. When the operating handle (1) is squeezed, it pierces the carbon dioxide cylinder and the carbon dioxide expands out, exerting pressure on the water.

❑ At the same time the seal (4) on the siphon tube (5) is opened and the high-pressure mixture of water and carbon dioxide can escape from the extinguisher to put out the fire.

How an aerosol works

❑ Aerosols are used to dispense many different products, such as hair spray, whipped cream, insecticides, air fresheners and paint.

❑ The diagram overleaf shows that an aerosol can contains two fluids, the product (which is what you want) and the propellant (in this case a compressed gas), which pushes the product out.

5 Solids, liquids and gases

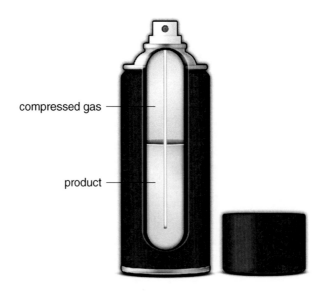

compressed gas

product

- The compressed gas is under great pressure and pushes down onto the product. When the aerosol is not in use, the spray nozzle seals the contents in the can, but when the nozzle is pressed down, the seal is broken and the pressure of the compressed gas forces the product out. The nozzle has a very small hole and the product is pushed out as a fine mist, called an aerosol cloud.

- It is recommended that aerosol cans are not stored at high temperatures. If the compressed gas heats up, the pressure increases even more and the can might explode.

- Some of the propellant escapes with the product and it disappears into the air. Manufacturers have to be careful that the propellants they use do not damage the ozone layer in the Earth's atmosphere, which protects us from harmful solar radiation.

Section 6 Magnetism and electromagnetism

Section 6a Units

- In this section you will come across the following units:
    - the **ampere** (A) is the unit of current
    - the **volt** (V) is the unit of voltage
    - the **watt** (W) is the unit of power
    - the **ohm** (Ω) is the unit of resistance.

Section 6b Magnetism

- A magnet exerts a force on other magnets and on magnetic material, such as iron.

- When you hold two magnets near to each other, you can feel them pull together (attract) or push each other away (repel). When you hold a magnet next to a piece of iron (or other magnetic material), it will attract the iron.

- The magnetic force of attraction or repulsion is experienced 'at a distance', i.e. the magnets do not have to be in contact. There is a region around the magnet in which the force is experienced.

- Properties of magnets:
    - they have north (N) and south (S) poles
    - like poles repel and unlike poles attract
    - they have a **magnetic field** around them – a region in space where their magnetism affects other magnets and magnetic materials.

- Magnetic materials contain iron, nickel or cobalt or compounds of iron. Steel is a magnetic material; it is an alloy containing iron. Aluminium, for example, is not a magnetic material.
  N.B. A magnetic material is **not** a magnet, but it can be made into a magnet. This is called **magnetising** the material.

Magnetic fields

❏ Magnetic fields can be described using **magnetic field lines** which show the direction and strength of the magnetic force.

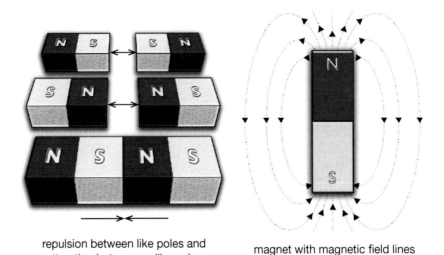

repulsion between like poles and attraction between unlike poles

magnet with magnetic field lines

- At any point in the field of a magnet, the direction of the field line is the direction of the force that would act on a N pole at that point.
- The magnetic field lines around a magnet point from the N pole of the magnet all the way back round to the S pole.
- The magnetic field is strongest where the field lines are closest or most dense.

❏ If two magnets are placed close together, their magnetic fields will interact. This interaction produces a force that causes the magnets to move if they are free to do so.

❏ You can show the pattern of magnetic field lines by sprinkling iron filings near the magnet. The iron filings arrange themselves in such a way as to show the magnetic field lines.

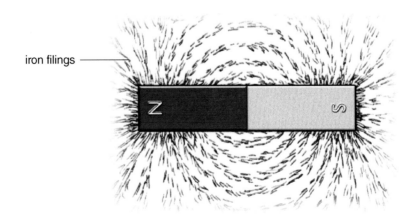

iron filings

6b Magnetism

❑ Magnetic field lines can be plotted, and their direction determined, using a plotting compass.

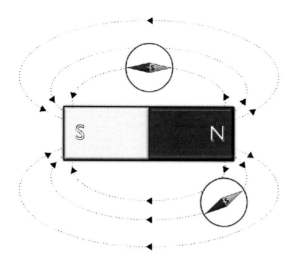

Method

1. Put the plotting compass next to one of the poles, and draw a pencil dot where the compass points.

2. Move the compass so that the tail end of the pointer aims at that dot.

3. Again, draw a pencil dot where the compass points. Repeat until the compass returns to the other pole.

4. Join up the dots.

5. Add arrows to show that the field points from the N pole to the S pole.

6. Repeat for another starting point.

> **Note**
>
> The field lines always have arrows from the N pole to the S pole and the lines never cross over each other or touch.

❑ If two magnets are placed close together, their magnetic fields will interact. The magnetic field pattern can be shown using iron filings or plotting compasses as above.

N.B. The same rules apply; the field lines go from the N pole to the S pole and never cross or touch.

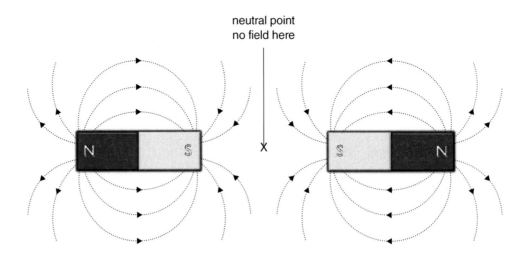

- In the diagram above the magnets are placed so that the poles repel. There will be a point between the magnets where the magnetic fields cancel each other out; this is called the **neutral point**.

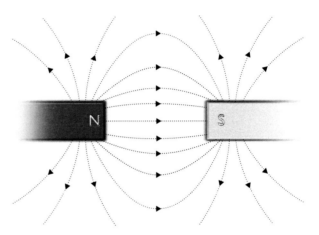

- In the diagram above the magnets are placed so that the poles are attracting. Between the magnets is a region where the field lines are nearly parallel. This means the field is non-uniform in that region because the magnets are weak.

- To create a uniform field, stronger magnets are required; we can use special magnets called magnadur magnets. Magnadur is a ceramic material containing oxides of iron used for making magnets. These magnets can be made with the poles on the faces of the magnets rather than on the ends. This means that when positioned with the opposite poles facing (see following diagram), the field lines will be straight and parallel **in the region between the poles**, i.e. the field is uniform in that region.

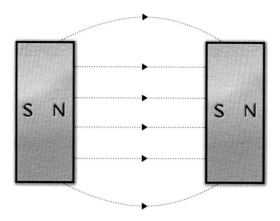

❏ Similarly, a horseshoe magnet has a uniform field in the region between its poles.

Test for magnetism

❏ A magnet exerts a force on other magnets as well as on magnetic materials. If you place a magnet next to an iron nail, it attracts the iron nail.

❏ If you place a bar magnet next to another bar magnet so that the opposite poles are close to each other, there will also be a force of attraction. The magnets will move towards each other.

❏ If you place a bar magnet next to an unidentified bar and the bar is attracted, you don't know whether the bar is a magnet or just made of magnetic material. But if you reverse the magnet and there is a force of repulsion, you know the unidentified bar is a magnet.

❏ The only **true test** for magnetism is **repulsion**.

6 Magnetism and electromagnetism

> **Note**
>
>
>
> Magnets always attract unmagnetised magnetic objects. Just because they are magnetic materials does not mean they are magnetised.
>
> The **test** for magnetism is **repulsion**; only two magnets will repel each other.

Hard and soft magnetic materials

- Magnetic materials are materials that can be magnetised, for example materials containing iron.

- Magnets can **induce magnetism** in other magnetic materials, i.e. make them into magnets.

- Each atom in a magnetic material acts as a small magnet. In an unmagnetised object, these atomic magnets line up with their neighbours to form aligned regions called domains, which point in random directions. In a magnetised object, the domains line up.

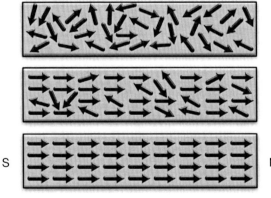

domains not lined up – unmagnetised

some domains lined up – weakly magnetised

all domains lined up – fully magnetised

- Steel is a **hard magnetic** material. It is difficult to magnetise, but once magnetised it is difficult to demagnetise. Steel is often used to make **permanent magnets**.

- Iron is a **soft magnetic** material. It is easy to magnetise, but also loses its magnetism easily. Iron is often used to make **temporary magnets**.

- The magnetic properties of steel make it a suitable choice to produce permanent magnets such as **bar magnets**.

- Bar magnets have numerous applications, including use as compass needles. They are always magnetised (permanent magnets) and **cannot be switched off**.

- The magnetic properties of iron make it a suitable choice for **electromagnets**. An electromagnet is a coil of wire with a soft iron core. When there is a current in the coil, the iron becomes magnetised. When the current is switched off, the iron loses its magnetism.

Section 6c   Electromagnetism

The magnetic effect of a current

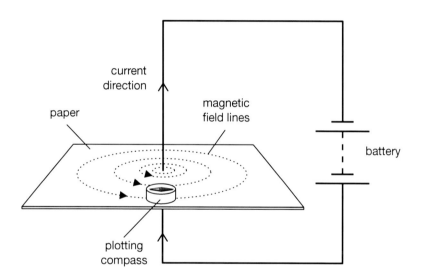

- A **magnetic field** is produced around a **current-carrying wire**. The field lines are **concentric circles** around the wire and can be shown using a plotting compass as in the diagram above. The magnetic field is strongest where the field lines are closest.

- The current produces a weak magnetic field if the current is small.

6 Magnetism and electromagnetism

❑ The magnetic field produced by a current-carrying wire has the following features.
- Increasing the current increases the strength of the magnetic field.
- The field is strongest closest to the wire and becomes weaker further away.
- Reversing the direction of the current reverses the direction of the magnetic field.

❑ The magnetic field produced by a single wire carrying a current is not very strong. A stronger field can be produced by a coil of wire.

○ The direction of the magnetic field around the wire can be found by using the **right hand grip rule**. Imagine gripping the wire in such a way that the thumb of your right hand points in the same direction as the current. Then your fingers curl around the wire in the direction of the magnetic field.

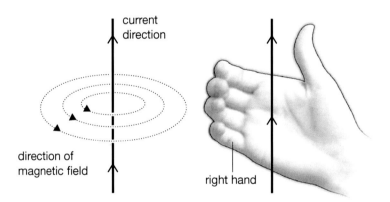

In this example, the **magnetic field** is in an **anti-clockwise** direction when viewed from above.

○ If the wire is formed into a flat coil, the magnetic field can be explored in a similar way, as shown in the following diagram.

6c Electromagnetism

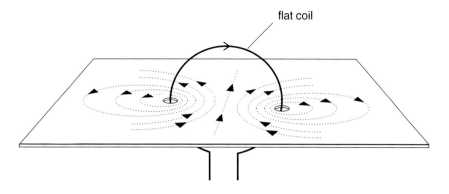

- A cylindrical coil of many turns of wire (a **solenoid**) is used to produce a stronger magnetic field.

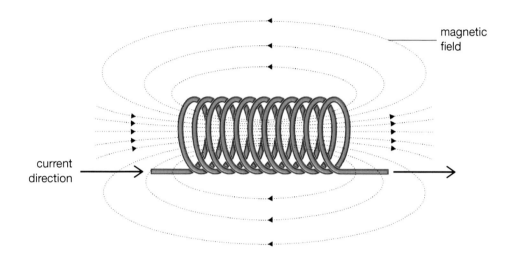

- The magnetic field produced by a current-carrying coil or solenoid has the following features:
  - the field is similar to the field of a bar magnet and there are magnetic poles at each end of the coil
  - the field inside the solenoid is nearly uniform
  - increasing the current increases the strength of the magnetic field
  - increasing the number of turns in the coil increases the strength of the magnetic field
  - having an iron core (for example by putting an iron nail in the coil as shown in the following diagram) greatly increases the strength of the magnetic field
  - reversing the direction of the current reverses the direction of the magnetic field
  - the field is strong close to the coil and weak further away.

  N.B. The magnetic field lines travel from S to N **inside** the solenoid.

6 Magnetism and electromagnetism

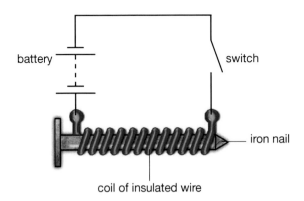

- A coil of wire with an iron core, connected into a circuit with a switch as above, is the basis of an **electromagnet**.

Practical applications of electromagnets

- Electromagnets also have numerous uses, such as in loudspeakers, large motors, generators, door locks and alarm bells. They can also be used to lift cars in scrap-yards and are used in the application of magnetic inks and paints. They use the magnetic effect of current and can **be switched on or off** (temporary magnets).

- A **relay** is a special type of switch turned on and off by an electromagnet.

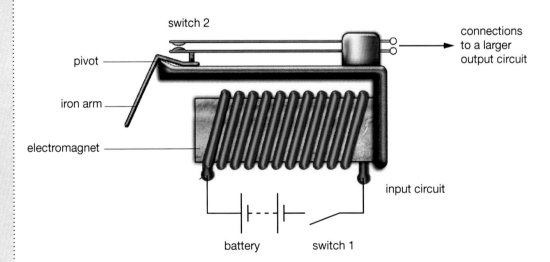

- When **switch 1** is **closed** in the diagram above, the input circuit is complete and there is a current in the circuit. The **electromagnet** becomes **magnetised** and it attracts the iron arm. The arm rotates about the pivot and pushes the two contacts of switch 2 together, switching on the output circuit. The output circuit is very often a more powerful circuit, such as the motor circuit for an elevator (lift). The rest of the output circuit is not shown in the diagram above.

○ The advantage of using a relay is that a **small current** in the input circuit can switch on a **large current** in the **output circuit**.

Force on a current-carrying conductor

❑ When a conductor carrying an electric **current** is placed in a **magnetic field** it will experience a **force**. That force will cause the conductor to move if it is free to do so.

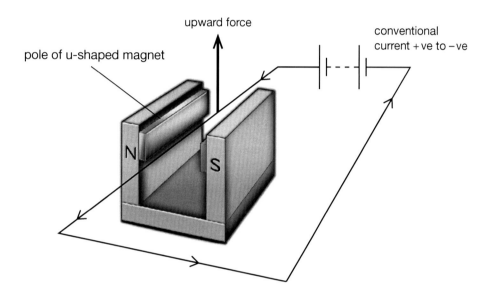

❑ The diagram shows a wire connected to a battery and placed between the poles of a magnet. When there is a current in the direction shown, the wire is seen to move upwards.

❑ This is because the current-carrying wire has its own magnetic field, which **interacts** with the field of the permanent magnet.

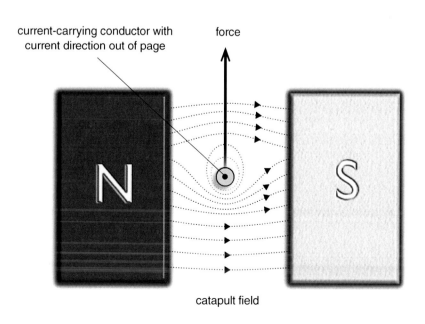

6 Magnetism and electromagnetism

- If the current-carrying wire is placed in a magnetic field (whose field lines are at right angles to the wire) then it will experience a force at right angles to both the current direction and the magnetic field lines.

- If either the direction of the current **or** the magnetic field is **reversed**, the direction of the **force** is **reversed**.

- If **both** the magnetic field and current are reversed, there is effectively no change in the force.

- The force can be **increased** by:
    - increasing the current in the wire
    - using a stronger magnet
      Remember: Stronger does not mean larger.
    - increasing the length of wire in the magnetic field.

> **Note**
>
> - If there is current in a wire in the presence of a magnetic field, a force is produced.
> - This is because the current-carrying wire has its own magnetic field, which interacts with the field of a permanent magnet.
> - But there is no force if the current direction and the magnetic field direction are parallel.

- This effect is used in motors (see pages 203-04) and loudspeakers (see page 205).

Fleming's left-hand rule

- Fleming's left-hand rule can be used to predict the direction of the force produced, for example, in d.c. motors.

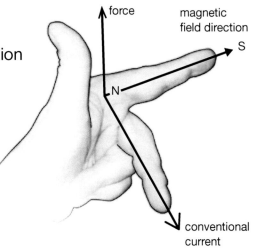

The directions of the current, magnetic field and force are related to each other as shown.
All three fingers have to be at **right angles** to one another.
A tip to help you remember what each finger represents is:

- **f**irst finger–**f**ield
- se**c**ond finger–**c**urrent
- **th**umb–**th**rust (i.e. force)

The d.c. motor

❑ A **d.c. motor** makes use of **direct current** (current in one direction).

❑ If there is a current in a coil placed between **fixed magnets**, forces act on the coil, producing a **turning effect**. There is a maximum force when the current is at right angles to the field. (There is zero force when the current is parallel to the field.)

❑ A d.c. motor transfers energy electrically into kinetic energy.

❑ Each end of the coil is connected to one half of a **split-ring commutator** against which conducting **carbon brushes** press. The coil is made of copper wire, as it is a good electrical conductor.

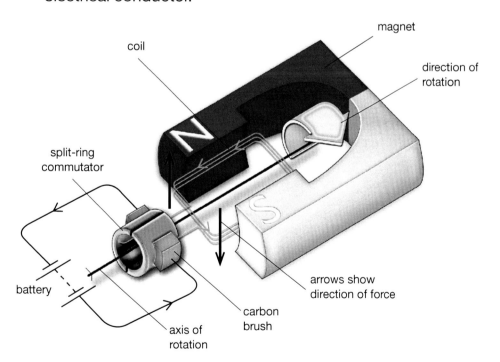

How it works

- The battery supplies a current to the coil via the split-ring commutator.
- Because the coil then has a current in it, it also has a magnetic field associated with it.
- This magnetic field interacts with the magnetic field of the permanent magnet.
- This causes an upward force on one side of the coil and a downward force on the other.
- The direction of each force is given by Fleming's left-hand rule.
- The split-ring commutator changes the direction of the current in the coil **every half-turn**, because as the coil rotates the commutator rotates.
- Therefore, the turning force keeps acting in the same direction as the coil rotates, keeping the coil rotating.
- The turning effect, and therefore the speed of revolution of the motor, can be increased by:
  - increasing the number of turns in the coil
  - using a stronger magnet
  - increasing the current.
- In an electric drill, for example, the rotating coil causes a drill bit to rotate, which drills a hole in wood, plastic, metal or stone.
- Many everyday applications use d.c. motors. They are used in washing machines, fans, air conditioning units and hair dryers.

Top Tip

How d.c motors work in 3 easy steps:

Step 1 When there is a current in a coil of wire, there is a magnetic field around that wire.

Step 2 This magnetic field interacts with the field of the permanent magnet.

Step 3 The interacting magnetic fields cause the coil to rotate.

The loudspeaker

❏ In a loudspeaker, energy is transferred electrically to kinetic energy, which is then transferred to the surroundings as sound.

❏ A loudspeaker contains an electromagnet and a permanent magnet connected to the speaker cone.

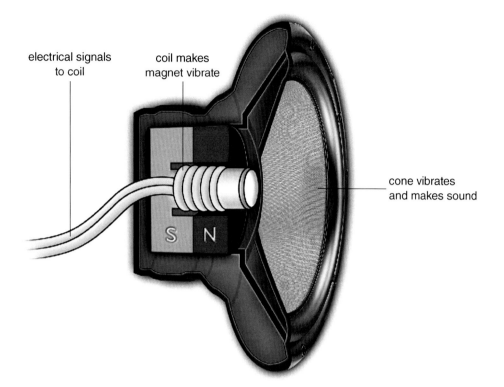

❏ The amplifier causes an alternating current in the coil of wire surrounding the electromagnet. Because the current is alternating, it produces a magnetic field in the electromagnet that is constantly changing direction.

❏ The changing magnetic field of the electromagnet interacts with the field of the permanent magnet, creating a force on the permanent magnet. This force is also constantly changing direction. As the magnet is attached to the speaker cone, the cone moves backwards and forwards, creating a sound wave as the air is alternately compressed and rarefied (see pages 112-13).

Forces on charged particles in a magnetic field

○ Current is the flow of charge and so it follows that there will be a force on any charged particle moving in a magnetic field (unless the motion is parallel to the field).

○ The diagram below shows an 'electron gun' inside a vacuum tube. A beam of electrons is fired towards the fluorescent screen at the front. A spot of light appears where the beam hits the screen.

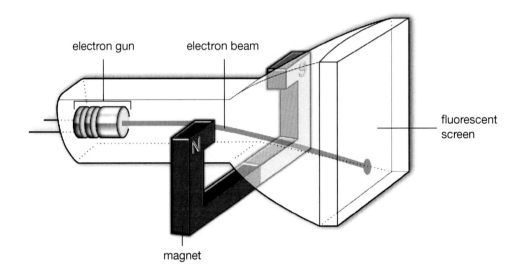

○ A magnet is placed against the outside of the tube and the spot on the screen moves downwards. This is because the direction of the magnetic field is from the N pole to the S pole and at right angles to the direction of movement of the electrons. The electrons are negatively charged, so as they move they form an electric current in the opposite direction to their motion (see page 63).

Applying Fleming's left-hand rule shows that the electrons experience a force downwards. Hence, the spot on the screen moves downwards.

Section 6d Electromagnetic induction

- Michael Faraday was the first person to generate electricity from a magnetic field using **electromagnetic induction**.

- When a wire is moved through a magnetic field, a small voltage is produced as the wire 'cuts' the magnetic field lines. This is known as an **induced voltage**. If the wire is part of a complete circuit, a current is produced in the wire by the induced voltage.

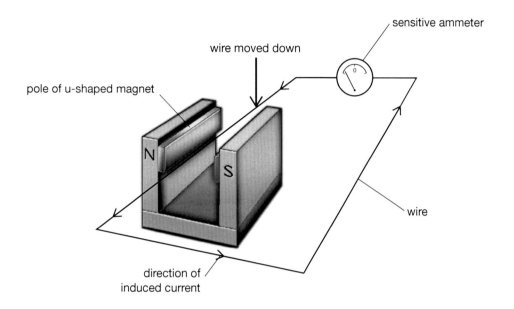

- The diagram above shows a wire being moved down between the poles of a magnet and 'cutting' the magnetic field lines. The same effect could be achieved by moving a magnet near a conductor. The sensitive ammeter will detect a current in the conductor if it is part of a complete circuit. The maximum current is observed when the wire cuts the magnetic field at right angles.

- The induced voltage (and hence the current) can be **increased** by:
    - moving the wire faster
    - using a stronger magnet to increase the magnetic field
    - increasing the length of wire cutting the magnetic field.

❑ When the induced voltage produces a current, its direction **opposes** the effect causing it. In other words, the magnetic field associated with the current caused by the induced voltage opposes the motion of the wire.

❑ There is **no voltage** and **no current** if the wire is **moved parallel** to the **magnetic field**. The wire must **cut** the magnetic field lines in order for a voltage or current to be induced.

Remember: The maximum effect is observed when the wire cuts the field at right angles.

Induced current in a coil

❑ Electromagnetic induction can also occur in a coil of wire when a magnet is pushed into and out of the coil, or the coil is moved towards and away from the magnet. In either case the magnetic field lines will be cut by the coil.

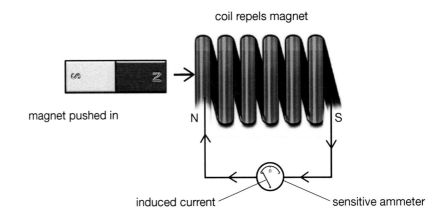

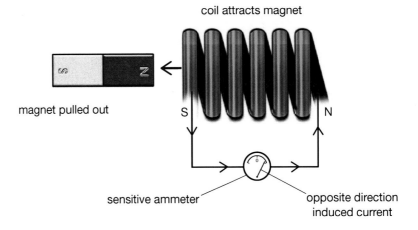

Remember: If the magnet is flipped round the current will be reversed.

6d Electromagnetic induction

- ❏ If the magnet is moved in and out several times, the needle on the sensitive ammeter will indicate that the current changes direction (or alternates) depending on the direction of movement.

- ❏ The induced voltage (and the current) can be **increased** by:
  - moving the magnet (or the coil) faster
  - using a stronger (not larger) magnet to increase the strength of the magnetic field
  - increasing the number of turns in the coil.

> **Note**
>
>
> - A conducting wire must cut the magnetic field lines in order for a voltage to be induced.
> - This voltage will be maximum when the field lines are cut at right angles.

- ❏ Electromagnetic induction is used for the commercial generation of electricity.

- ❏ It is not practical to move a magnet backwards and forwards in a coil and so the motion is achieved by spinning a coil of wire inside a magnetic field, or by spinning a magnet inside a coil of wire.

- ❏ A bicycle dynamo is a simple example of a practical device for generating electricity. The movement of the bicycle wheel spins a magnet in a coil of wire.

- ❏ A turbine is a larger-scale machine whose blades transfer the kinetic energy from a moving fluid such as steam, wind, water, etc. (see pages 148-50). The blades rotate and translate this motion to move a magnet relative to a coil of wire so inducing a current.

The a.c. generator

❑ An **a.c. generator** produces **alternating** (backwards and forwards) voltage and current.

Remember: The term a.c. stands for 'alternating current'.

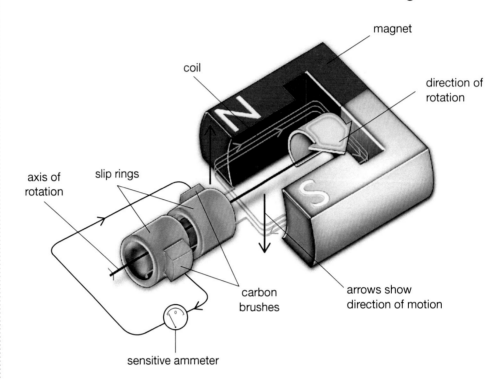

❑ In a generator, kinetic energy is transferred electrically to the external circuit.

❑ Generators may use a **rotating coil** between **fixed magnets**, as shown in the diagram above.

❑ Each end of the coil is connected to a **slip ring** against which conducting **carbon brushes** press. The coil is made from copper wire, as copper is a good electrical conductor.

❑ The maximum voltage and current can be increased by:
- increasing the number of turns in the coil
- using a stronger (not larger) magnet
- rotating the coil faster.

❑ The diagram opposite shows a graph of voltage output against time, and how the output relates to the position of the coil in the magnetic field.

6d Electromagnetic induction

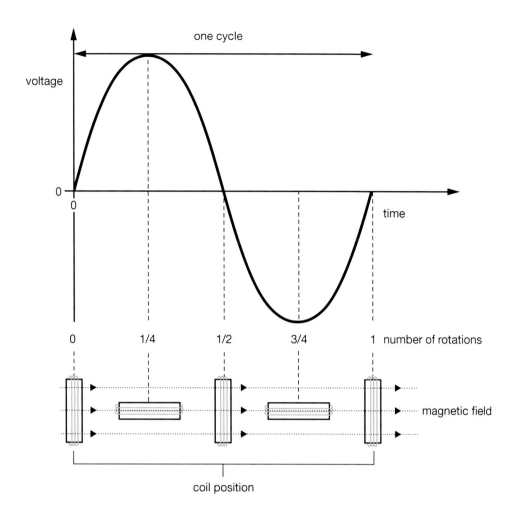

How it works

☐ The coil is rotated in the magnetic field, **cutting** the magnetic field lines so that an alternating voltage is induced.

☐ Because one side of the coil is always connected to the same slip ring, as the coil rotates, the current in the circuit changes direction every half-cycle. This produces a.c. (alternating current) in an external circuit.

☐ The voltage and the current are at a **maximum** when the coil is **horizontal** (parallel to the field) as **two sides of the coil** are then **cutting** the field lines at the greatest rate (at **right angles**).

☐ The voltage and the current are **zero** when the coil is **vertical** (at right angles to the field) as no field lines are being cut because **two sides of the coil** are then moving **parallel** to the field lines.

☐ The same effect is produced if a magnet rotates within a static coil of wire.

Transformers

○ A transformer is used to **change the size** of an **alternating voltage**.

○ It is made by winding two insulated coils around a soft iron core, as shown below. These coils are known as the **primary (input)** and the **secondary (output)** coils. (There are many more turns on the coils of a real transformer.)

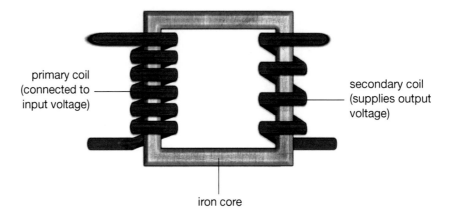

primary coil (connected to input voltage)

secondary coil (supplies output voltage)

iron core

○ When there is an alternating current in the **primary** (input) coil, it sets up an **alternating magnetic field** in the primary coil and in the iron core.

○ This changing magnetic field is transferred through the core and into the **secondary** (output) coil.

○ This induces an **alternating voltage** in the secondary (output) coil.

Top Tip

How transformers work in three easy steps:

Step 1 An **alternating current** in the primary coil sets up an **alternating magnetic field** in the primary coil.

Step 2 This alternating magnetic field is **transferred** through the iron core.

Step 3 The **alternating magnetic field** cuts through the **secondary coil** and **induces** an **alternating voltage**.

Step-up and step-down transformers

- There are **more secondary turns** than primary turns in a **step-up** transformer.
  The **secondary voltage is larger** than the primary voltage, i.e. it is stepped up.

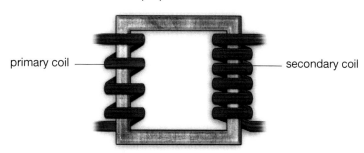

step-up transformer

primary coil — secondary coil

- There are **fewer secondary turns** than primary turns in a **step-down** transformer.
  The **secondary voltage is smaller** than the primary voltage, i.e. it is stepped down.

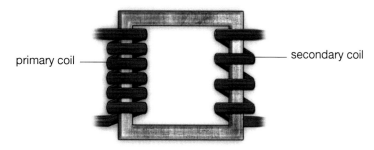

step-down transformer

primary coil — secondary coil

- The number of turns and voltage in the primary and secondary coil are related by the following expression:

$$\frac{\text{input (primary) voltage}}{\text{output (secondary) voltage}} = \frac{\text{primary turns}}{\text{secondary turns}}$$

$$\boxed{\frac{V_p}{V_s} = \frac{N_p}{N_s}}$$

V_p = primary voltage (V)
V_s = secondary voltage (V)
N_p = number of turns on primary coil
N_s = number of turns on secondary coil

6 Magnetism and electromagnetism

Example 6d (i)

A transformer transforms 240V a.c. to 12V a.c. for a model car racing set. The secondary coil has 50 turns.

(i) State the equation linking the primary and secondary voltages with the primary and secondary turns

(ii) Calculate the number of turns on the primary coil.

Answer

(i) $$\frac{\text{input (primary) voltage}}{\text{output (secondary) voltage}} = \frac{\text{primary turns}}{\text{secondary turns}} \quad \text{or} \quad \frac{V_p}{V_s} = \frac{N_p}{N_s}$$

(ii) **Step 1** List the information in symbol form and change into appropriate and consistent SI units if required.

$$V_p = 240\,V$$
$$V_s = 12\,V$$
$$N_s = 50$$
$$N_p = ?$$

Remember: There are no units for the number of turns.

Step 2 Rearrange the equation.

$$\frac{V_p}{V_s} = \frac{N_p}{N_s} \quad \Rightarrow \quad N_p = \frac{V_p \times N_s}{V_s}$$

Step 3 Calculate the answer by putting the numbers into the equation.

$$N_p = \frac{240 \times 50}{12} = 1000$$

Efficiency of transformers

○ Transformers are **very efficient**; their power output is almost as high as their power input.

○ If a transformer could be made **100% efficient** then the power leaving the transformer would equal the power coming into the transformer.
Remember: Power $P = I \times V$ (see page 55).

For 100% efficiency:
input power = output power

$$\frac{\text{voltage in}}{\text{primary}} \times \frac{\text{current in}}{\text{primary}} = \frac{\text{voltage in}}{\text{secondary}} \times \frac{\text{current in}}{\text{secondary}}$$

$$\boxed{V_p \times I_p = V_s \times I_s}$$

V_p = primary voltage (V)
I_p = primary current (A)
V_s = secondary voltage (V)
I_s = secondary current (A)

N.B. If V_s is greater than V_p, as in a step-up transformer, then I_s must be less than I_p.

○ **No device is 100% efficient** and the energy output is always less than the energy input. Even though transformers are not 100% efficient, we often assume that they are in calculations.

○ One reason why transformers are not 100% efficient is because the **resistance** in both the primary and secondary coils causes heating.

○ Their efficiency might also be affected by the primary coil's magnetic field not linking the secondary coil with maximum effect.

Transmission of electricity

○ Power supplied to our homes is generated in a power station.

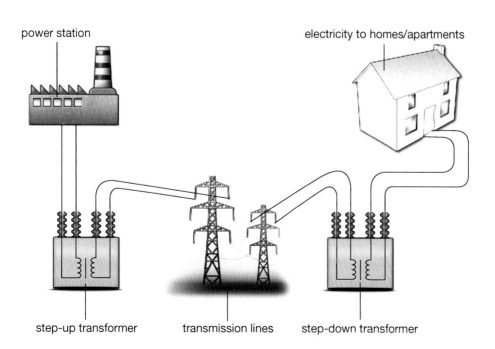

6 Magnetism and electromagnetism

- **Step-up** transformers are used to **increase the voltage** and to **decrease the current** in cables used in transmission lines.

- The current is decreased deliberately in order to reduce the power 'loss' in the cables (when there is a current in the cables, they heat up and energy is wasted).

    The power 'loss' in the cables can be calculated using the following equation.

    $$P = I \times V = I \times (I \times R) = I^2 R$$

    P = power (W)
    I = current (A)
    V = voltage (V)
    R = resistance (Ω)

    N.B. A smaller current will give a smaller power 'loss'.

- So, by using step-up transformers to reduce the current, the **heating effect is minimised**. This is an advantage as it means cheaper and thinner cables can be used in transmission lines.

- **Step-down** transformers are used to **reduce the voltage** before the electricity enters our homes.

- Transformers **only** work using **a.c.** and so that is why the current entering our homes is **a.c.** rather than **d.c.**

> **Note**
>
>
> **Remember:** Transformers will only work if there is an **alternating magnetic field**. This alternating magnetic field is only present if there is an **alternating current**. Therefore transformers will only work using a.c. and not d.c.

Section 7   Radioactivity and particles

Section 7a   Units

- In this section you will come across the following units:
  - the **becquerel** (Bq) is the unit of activity of a radioactive source
  - the **centimetre** (cm) is a unit of length
  - the **hour** (h), the **minute** (min) and the **second** (s) are units of time.

Section 7b   Radioactivity

The nuclear atom

- This diagram is a simple representation of atomic structure.

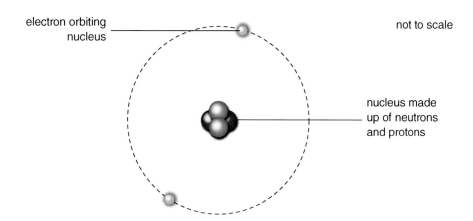

- The nucleus of an atom contains positive particles called **protons** and neutral (uncharged) particles called **neutrons**. The 'empty space' of the atom around the nucleus contains tiny negatively charged **electrons** in orbits or 'shells'.

- The number of protons in the nucleus is known as the **proton number Z** or **atomic number**. In a **neutral** (not ionised) atom, the atomic number is also equal to the number of electrons orbiting the nucleus.

- The **nucleon number A** or **mass number** is the number of **nucleons** (protons and neutrons) in the nucleus.

- This structure of an atom is often written in the following format, known as nuclide notation:

  $^{A}_{Z}X$ where **X** is the **chemical symbol** for the element.

For example, $^{1}_{1}H$ (hydrogen-1), $^{16}_{8}O$ (oxygen-16) and $^{24}_{12}Mg$ (magnesium-24).

- $^{1}_{1}H$ has one proton and no (1−1) neutrons in its nucleus and one electron orbiting the nucleus.
- $^{16}_{8}O$ has 8 protons and 8 (16−8) neutrons in its nucleus and 16 electrons orbiting the nucleus.
- $^{24}_{12}Mg$ has 12 protons, and 12 (24−12) neutrons in its nucleus and 24 electrons orbiting the nucleus.

❏ Each **element** has a different **atomic number**, and the elements are arranged in order of **increasing** atomic number in the periodic table.

❏ The masses and charges of the subatomic particles are given in this table in terms of the mass and charge of the proton.

Particle	Relative mass	Relative charge	Symbol
proton	1	+1	$^{1}_{1}p$
neutron	1	0	$^{1}_{0}n$
electron	1/2000	−1	$^{0}_{-1}e$

Isotopes

❏ Nuclei with the same **atomic number** can have different **mass numbers**.
For example: $^{12}_{6}C$ $^{14}_{6}C$

These nuclei are called carbon-12 and carbon-14. They are different **isotopes** of the element carbon.

❏ Isotopes of an element have the same number of **protons** but a different number of **neutrons**.

❏ If the nucleus of an atom has too many or too few neutrons compared to the number of protons, it becomes unstable and is called a **radioactive isotope**.

❏ The isotope carbon-14 is radioactive and can be used in **carbon dating** (see page 227).

Characteristics of radioactive emissions

❑ Many nuclei of elements in the periodic table are **stable** but some are **unstable**. Unstable nuclei undergo **radioactive decay**.

❑ When a radioactive nucleus decays, it may emit one or more of the following types of radiation: **alpha (α) particles**, **beta (β⁻) particles** or **gamma (γ) rays**.

❑ Alpha particles are stopped by paper, beta particles are stopped by 3mm of aluminium and gamma rays are greatly reduced in intensity by lead.

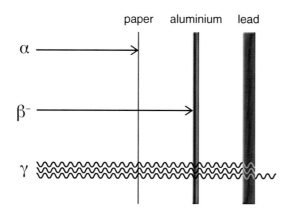

❑ The radiations alpha, beta and gamma have their own individual characteristics, as shown in the table below.

	Alpha particle	Beta particle	Gamma ray
Nature	2 protons and 2 neutrons (helium nucleus)	electron	electromagnetic radiation
Symbol	$^4_2\alpha$ or ^4_2He	$^0_{-1}\beta^-$ or $^0_{-1}e$	γ
Charge	positive	negative	uncharged
Affected by magnetic and electric fields	yes	yes	no
Penetrating power	weak – stopped by thin paper or approximately 6 cm of air	moderate – stopped by a few mm of aluminium	strong – only stopped by many cm of lead or many m of concrete
Relative ionising effect	strongest	medium	weakest
Dangerous	yes	yes	yes

7 Radioactivity and particles

- There is always radiation present all around us. This is known as **background radiation**. Background radiation comes from natural sources.
    - **Cosmic rays** – A constant stream of radiation comes to the Earth from the stars and the Sun; much, but not all, of the radiation is absorbed by the Earth's atmosphere.
    - **Rocks and soils** – Radioactive materials exist naturally in rocks and soils and give off radioactive radon gas, which we inhale.
    - **Organic material** – All plants and animals contain tiny amounts of radioactive carbon and potassium, which we eat.
    - Some background radiation comes from **human-made sources**, including medical procedures (such as x-rays and radiotherapy), nuclear power stations (including nuclear waste) and nuclear weapons testing.

- **Ionising radiation** is powerful enough to knock electrons from atoms, leaving them positively charged. These positively charged particles are called positive **ions**. Alpha, beta and gamma radiation all have an ionising effect, to some degree (see the table on the previous page).

Harmful effects of the different kinds of emission

- Radioactive emissions can ionise the atoms of the material they travel through. This may cause chemical reactions to occur. This is particularly dangerous if these reactions occur within the living cells of humans. Radiation can damage cells and tissues, and cause mutations (changes in genetic material).

- α-particles are the most ionising, so α-radiation is potentially the most harmful if ingested into our bodies. However, α-particles are not very penetrating and so cannot get past our skin when produced outside the body. Consequently, α-radiation is more dangerous within our bodies than on the outside of our bodies.

- β^--particles and γ-rays are weaker at ionising if ingested into our bodies but can penetrate through our skin more easily when travelling through air. This is what makes them dangerous from outside the body.

- **Safety precautions** for storing and handling include, but are not limited to:
  - **time**, i.e. reducing the amount of time a person is exposed to radiation
  - wearing a film badge that alerts the wearer to the possibility of dangerous levels of radiation
  - **distance**, i.e. using forceps or robotic manipulators to hold radioactive sources
  - **shielding**, i.e. storing radioactive materials in thick lead containers
  - wearing lead-lined clothing and gloves
  - working behind a lead-glass shield or putting a wall between the person and the source.

Detection of ionising radiation

- All radiation produced by radioactivity can be detected using **photographic film** or a **Geiger–Müller tube** as shown in the following experiment.

To investigate the penetrating power of radiation from different radioactive materials

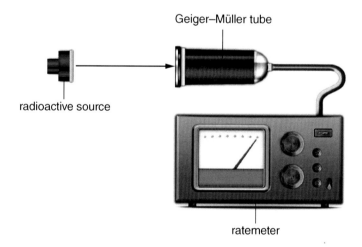

- The experiment, conducted by the teacher, is as follows:

Method

1. Use the Geiger–Müller tube and ratemeter to measure background radiation. This level of activity should be subtracted from all readings taken to give the activity due to the source alone.

2. Set up the apparatus as shown in the previous diagram with the alpha source a small fixed distance from the Geiger–Müller tube, and take and record a reading of the activity.

3. Insert a sheet of paper between the source and the Geiger–Müller tube and take and record a reading of the activity.

4. Replace the paper with a sheet of aluminium and then with a sheet of lead, recording the activity with each.

5. Repeat the process in steps 2 to 4 with the beta source and then with the gamma source, making sure the distance of the source is the same in each case.

6. Construct a table of results showing the activity recorded in each case.
Remember: Subtract the background count from the activity measured in each of your results.

❏ Your results should confirm that α-particles are stopped by paper, β⁻-particles are stopped by 3 mm of aluminium and γ-rays are greatly reduced in intensity by lead.

Example 7b (i)
A student designs an experiment to investigate what type of radiation a radioactive isotope emits, as shown below. The ratemeter measures radiation received in counts/minute.

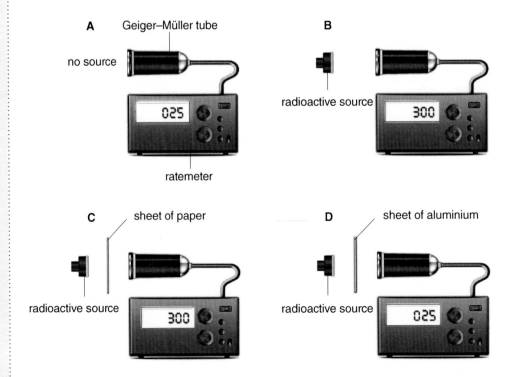

N.B. The SI unit of activity is the Bq, which is the number of decays per second (see page 229), but for school lab sources the count rates are so small it's sensible to work in counts/minute.

The experiment was then carried out by the teacher and the following measurements were taken:

A Detector and no source, giving a background reading of 25 counts/minute.

B Detector and radioactive source, giving a reading of 300 counts/minute.

C Detector, source and paper, giving a reading of 300 counts/minute.

D Detector, source and aluminium sheet, giving a reading of 25 counts/minute.

The student concludes that only β^--particles are emitted by the source. Explain how the results obtained show that only β^--particles are emitted.

Answer

When paper was placed between the source and the detector in diagram **C**, the reading did not change, so there can be no α-radiation emitted. This would have been stopped by paper.

When the aluminium was placed between the source and the detector in diagram **D**, the reading dropped to the background value (the reading in **A**), showing that the radiation was stopped by the aluminium. Therefore the radiation must be β^--particles.

There can be no γ-rays present. If there were, they would not be stopped by the aluminium, so the reading would not drop in diagram **D** to the background value.

Effects of magnetic fields and charged plates

- ❏ α-particles are positively charged and β^--particles are negatively charged. Because they are charged, they are affected by magnetic fields and when they pass between charged plates.

- ❏ γ-rays have no charge and therefore are not influenced by a magnetic field or when they pass between charged plates.

❑ The diagram below shows how the three kinds of radiation behave when they pass between charged plates.

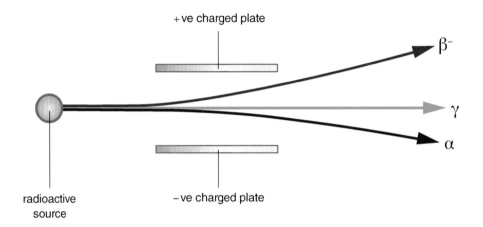

❑ α-particles are **deflected less** than β⁻-particles because they have a much **greater mass** and therefore need a much **bigger force** to deflect them.

❑ The diagram below shows how the three kinds of radiation behave in a **magnetic field**.

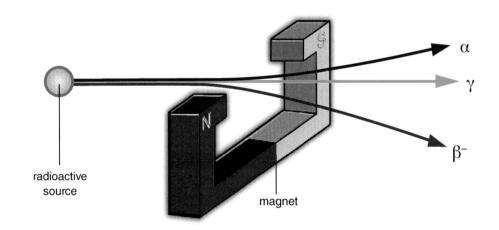

❑ The direction of deflection of α- and β⁻ particles in a magnetic field can be found by using **Fleming's left-hand rule** (see pages 202-03).

Remember: Fleming's left-hand rule is based on conventional current (positive to negative).

Contamination versus irradiation

❏ When a radioactive material is transferred from one environment to another, the effects of the radioactive material can **contaminate** the second environment. For example, when the Fukushima nuclear power station exploded in Japan in 2011 following the earthquake and tsunami, radioactive material spilled out and contaminated the surrounding areas. About 800 km^2 of land has been declared unfit for human habitation. It will remain radioactive until the activity has reduced to a safe level, which will be many years (see pages 229-30).

❏ If somebody breathes in radioactive gas, or swallows radioactive material, they become contaminated. The radioactive material remains in their body until its activity reduces to a safe level.

❏ Radioactive waste must be disposed of carefully to avoid dangerous contamination. Waste is produced by nuclear power stations, and also by other industries and businesses, including hospitals (which use radioactive materials for medical purposes) and schools (which use radioactive materials for educational purposes). There are strict regulations regarding the **disposal** of such materials, and the methods of disposal depend on the level of the activity of the radioactive material.

- High-level waste, such as the waste products from nuclear power stations, is sealed in glass and steel and stored deep underground.

- Low-level waste consists of radioactive materials that do not remain radioactive for very long (they have very short half-lives), (see pages 229-30). It is stored in a special facility until it is no longer a danger and then disposed of with normal waste products.

❏ If a substance is exposed to radiation, it is said to be **irradiated**. Irradiation by ionising radiation can cause harm, but it can also be useful.

- When something has been irradiated by γ-rays, for example, the irradiation stops as soon as the source of γ-rays has been removed. Irradiation by γ-rays is used to sterilise many products, including medical equipment, because they kill bacteria. The medical equipment does **not** become radioactive just because it has been irradiated.

Uses of radioactive isotopes

- Radiation is used to sterilise medical equipment, as described above.

- Irradiating food products can prolong the life of the food by killing the micro-organisms that cause food to decompose.

- β⁻-particles can be used to monitor the thickness of paper. In this thickness-monitoring process, the number of β⁻-particles that pass through the paper is inversely related to the thickness of the material. If the paper becomes too thick, fewer β⁻-particles are detected. There is an automatic feedback system that will adjust the paper thickness accordingly.

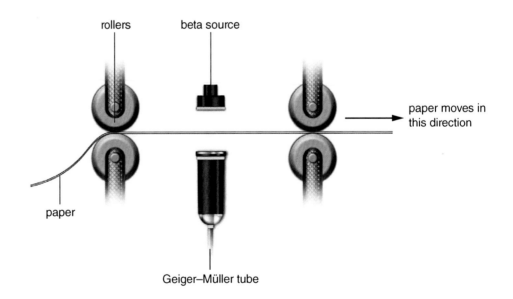

α-particles would not pass through at all and γ-radiation would pass through unimpeded. So neither α-particles nor γ-rays would be useful in this process.

- ☐ **Medical tracers** are used to detect blockages and other problems in vital organs. A small amount of radioactive isotope is injected into a patient's bloodstream. Such isotopes emit γ-radiation, which will pass through the body to an external imager. The imager can follow the path the isotope takes and detect where there is evidence of poor blood supply and a possible blockage. The isotopes have short half-lives, so they are quickly eliminated from the body.

- ☐ γ-rays can be used in **radiotherapy**; beams of γ-radiation are fired directly at **cancer cells** to kill them.

- ☐ An isotope of carbon is used in **radioactive carbon dating**. All living organisms contain a small amount of carbon-14, which is radioactive, with a half-life of 5700 years. When the organism dies, the remaining carbon-14 decays slowly. The ratio of carbon-14 to the non-radioactive carbon-12 can be used to calculate when the organism was last living. Carbon dating is therefore very useful in archaeology.

Radioactive decay

- ☐ Unstable elements undergo the random process of radioactive decay. The nucleus of an unstable atom breaks up to form a different nucleus (i.e. a different element) and releases energy.

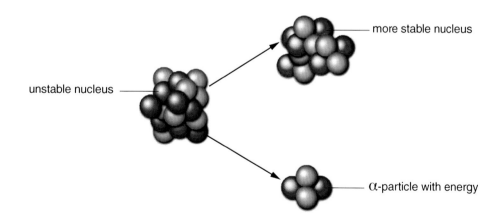

- ☐ An α-particle is identical to a helium nucleus, so it has an **atomic number of 2** and a **mass number of 4**.

7 Radioactivity and particles

- In α-decay a nucleus loses **two protons** and **two neutrons**, so the mass number is reduced by 4 and the atomic number by 2. For example, using nuclide notation:

$$^{226}_{88}\text{Ra} \rightarrow \, ^{222}_{86}\text{Rn} + \, ^{4}_{2}\alpha$$

The mass numbers and proton numbers balance on both sides of the equation:

$226 = 222 + 4$
$88 = 86 + 2$

- In β-decay a **neutron changes** into a **proton** and an **electron**. The atomic number of the nucleus therefore increases by 1 and the mass number stays the same. The new proton stays in the nucleus but the electron is expelled as a β-particle. For example, using nuclide notation:

$$^{131}_{53}\text{I} \rightarrow \, ^{131}_{54}\text{Xe} + \, ^{0}_{-1}\beta^{-}$$

The mass numbers and proton numbers balance on both sides of the equation:

$131 = 131 + 0$
$53 = 54 - 1$

- A β-particle is a high-energy (high-speed) electron emitted from the nucleus.

- γ-rays are emitted when a nucleus decays but is still in a slightly unstable state after emission of α or β particles. γ-emission causes **no change** in atomic or mass number.

- Neutron emission is a process by which an unstable nucleus can emit a neutron to produce a more stable nucleus. In this case the atomic number stays the same because the number of protons stays the same, although the mass number decreases by 1. The nucleus becomes a different isotope of the same element.

$$^{13}_{4}\text{Be} \rightarrow \, ^{12}_{4}\text{Be} + \, ^{1}_{0}\text{n}$$

Again, the mass numbers and proton numbers balance on both sides of the equation:

$13 = 12 + 1$
$4 = 4 + 0$

Half-life

- The **activity** of any radioactive source is measured in a unit called the **becquerel (Bq)**.

  An activity of **1 Bq is one nucleus decaying per second**. If a source has an activity of 100 Bq, it follows that 100 radioactive nuclei are decaying per second.

- The activity of radioactive sources **decreases with time**.

- Some radioactive sources are **more unstable** than others and decay at a **faster rate**.

- **Remember:** 'Decay' does not mean that radioactive material disappears. The unstable nuclei of the material change to stable nuclei of a different element.

- The bigger the mass of a given source, the greater the activity since the activity is directly proportional to the number of radioactive nuclei remaining.

- Radioactive decay is **not affected** by temperature or pressure; it is spontaneous, i.e. it happens on its own.

- Radioactive emission is a completely **random** process. It is impossible to predict when a particular nucleus will decay.

- However, the average time taken for **half** of the unstable nuclei in a sample of a particular radioactive isotope to decay is **always the same**. This time is known as the **half-life**.

- The half-life is also the time taken for the **activity** of a radioactive isotope to **drop by half** of its original value.

- Half-life is different for different radioactive isotopes. For example, the half life of uranium-234 is 245 000 years but the half-life of radon-222 is only 3.82 days. The half-lives of some radioactive elements are very small fractions of a second.

- The **count rate** detected due to a source is proportional to the activity of the source.

❑ When calculating the half-life, the count rate must be corrected to account for **background radiation**. You must subtract the background count rate from the measured count rate before you start performing calculations.

> **Note**
>
>
> Although activity is measured in Bq (counts per second), counters are often calibrated in counts/minute. It is acceptable to work in either unit, but never use both together in the same calculation.

❑ The graph below shows a typical example of a radioactive isotope decaying over a period of time.

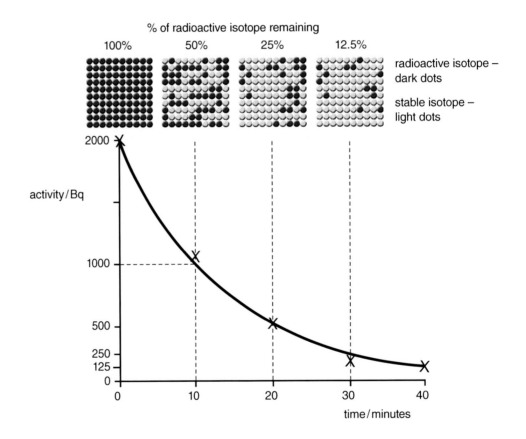

- ❑ Since the decay of an isotope is random, the curve is actually a curve of **best fit**.

- ❑ If values of activity in Bq (or count rate in counts/s or counts/min) are plotted against time in minutes, the value of the half-life can be found from the previous graph. Suppose the initial reading (100% in the graph) is 2000 Bq. When the activity falls to 1000 Bq, one half-life has elapsed. By drawing a horizontal line from the 1000 Bq point on the activity axis across to the curve, and then from there down vertically to the time axis, the half-life can be measured. In this case it is 10 minutes.

- ❑ To calculate the half-life of an isotope without a graph, the count rate before and after and the total time elapsed are needed.

Example 7b (ii)
In an experiment, a radioactive isotope's activity falls from 200 Bq to 25 Bq in 75 minutes. Calculate its half-life.

Answer
200 Bq ➡ 100 Bq ➡ 50 Bq ➡ 25 Bq

The activity has halved three times; therefore three half-lives have elapsed.

half-life = 75 minutes ÷ 3
half-life = 25 minutes

- ❑ To calculate the **activity** of an isotope after a period of time, the half-life, the starting activity and the time elapsed are needed.

Example 7b (iii)
The half-life of a substance is 3.0 days. The initial count rate recorded next to a sample of this substance is 2050 counts/minute and the background radiation count rate is 50 counts/minute. Calculate the count rate 9.0 days later.

Answer

The actual initial count rate is 2000 counts/minute because background is 50 counts/minute.

The half-life is 3.0 days; therefore after 9.0 days three half-lives have elapsed.

2000 counts/minute ➡ 1000 counts/minute ➡ 500 counts/minute ➡ 250 counts/minute

Therefore the activity is 250 counts/minute 9.0 days later, but the background radiation count rate of 50 counts/minute is assumed constant, so the actual count rate is 300 counts/minute.

Section 7c Fission and fusion

- Radioactivity, nuclear fission and nuclear fusion are three types of nuclear reaction. They all result in a release of energy.

Fission

- Nuclear power reactors (for example, at nuclear power stations) use a nuclear reaction called **nuclear fission**. Fission means splitting.

- Fission occurs when an unstable large nucleus splits to form two smaller nuclei, with the release of energy.

- Two isotopes commonly used as nuclear fuels are uranium-235 and plutonium-239. Both these isotopes are large nuclei and can be split relatively easily, especially when neutrons collide with them.

- Naturally occurring uranium is mostly uranium-238, with a very small percentage of the isotope uranium-235. When a neutron strikes (collides with) a uranium-235 nucleus, the nucleus becomes unstable (uranium-236).

$$^{235}_{92}U + ^{1}_{0}n \rightarrow ^{236}_{92}U$$

- The unstable uranium-236 splits into two smaller radioactive nuclei, called **daughter nuclei**.

7c Fission and fusion

- Neutrons are released as well as energy. This energy appears as kinetic energy of the neutrons and γ-radiation. The energy transfers into thermal energy when it is absorbed by the surrounding materials in the reactor.

- Neutrons produced from one fission strike other uranium-235 nuclei, causing further fissions and releasing even more neutrons, and a **chain reaction** can develop. The chain reaction could cause an atomic explosion because of the huge amount of energy that is released.

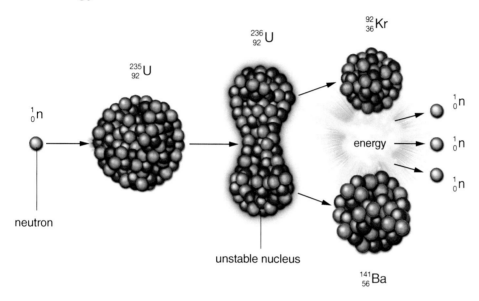

- The equation for this reaction can be expressed as:

$$^{1}_{0}n + ^{235}_{92}U \rightarrow ^{141}_{56}Ba + ^{92}_{36}Kr + \left(3 \times ^{1}_{0}n\right) + \text{ENERGY}$$

- In a nuclear reactor this chain reaction is **controlled** to release the energy at a steady rate and to stop the reaction from going too fast. Control rods and a moderator work together to control the reaction, but they have different functions.

- The moderator keeps the reaction going and the control rods prevent the reaction going out of control.

- The **moderator**, often made of graphite, **slows down the neutrons** released from the chain reaction so that they interact with the uranium fuel. If the neutrons are moving too fast, then they do not get captured by the nucleus of a uranium atom and the chain reaction stops. Ideally one neutron from each fission must go on to strike another nucleus for the process to keep going.

- ❑ The moderator does not provide any protection if the reaction goes out of control. Control rods are inserted between the fuel rods for this purpose.

- ❑ The **control rods** are usually made of **boron**, which is stable and won't be affected by the bombarding neutrons. The boron **absorbs the neutrons** and so reduces the chain reaction by preventing them from striking uranium nuclei. The rate of fission can be controlled by raising or lowering the boron control rods between the uranium fuel rods.

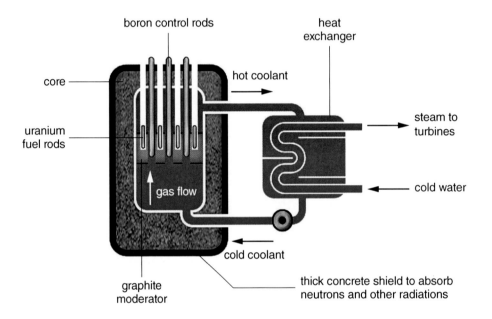

- ❑ The **coolant** (usually water or carbon dioxide) has two functions in a nuclear power station. It is circulated through the reactor core to absorb the thermal energy and cool down the reactor, and it is then passed through a heat exchanger to boil water and create steam, which drives turbine generators.

- ❑ The energy released in a nuclear reactor is far greater than that released by burning fossil fuels. This energy comprises the kinetic energy of the neutrons, not all of which are captured by the boron control rods, and γ-radiation, which can cause a serious health hazard.

- ❑ Radiation leakage would be very harmful not only to the people working in the nuclear power station but also the nearby environment.

- ❏ **Shielding** is provided by covering the reactor with a large mass of absorbent material (typically a combination of steel plate and a thick layer of concrete) designed to reduce the radiation to a safe level.

Fusion

- ❏ **Nuclear fusion** is the process that occurs in the Sun and other stars to release energy. Fusion means combining.

- ❏ Fusion occurs when two small stable nuclei combine together to form a larger stable nucleus, with a release of energy.

- ❏ In one simple fusion reaction, deuterium ($^{2}_{1}H$) and tritium ($^{3}_{1}H$) (isotopes of hydrogen) nuclei combine to form helium nuclei and neutrons resulting in an overall loss of mass (the mass of the neutron and helium is less than the combined mass of the deuterium and tritium) and a release of energy.

$$^{2}_{1}H + ^{3}_{1}H \rightarrow ^{4}_{2}He + ^{1}_{0}n + \text{ENERGY}$$

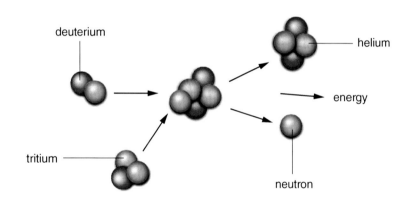

- ❏ The nuclei have to get very close to collide, but the deuterium and tritium nuclei are both positively charged, as all nuclei have protons in them, and so there is electrostatic repulsion between them. As a result, fusion does not happen at low temperatures and pressures.

- ❏ The nuclei must have enough kinetic energy to overcome this electrostatic repulsion and allow the positive nuclei to get close enough for fusion to occur. This requires extremely high temperature and high pressure, such as occur in the core of stars. The temperature at the core of the Sun can reach $1.5 \times 10^{7}\,°C$.

Section 8 Astrophysics

Section 8a Units

❑ In this section you will come across the following units:
- the **kilogram** (kg) is the unit of mass
- the **metre** (m) is the unit of length
- the **second** (s) is the unit of time
- the **metre per second** (m/s) is the unit of speed
- the **metre per second squared** (m/s^2) is the unit of acceleration
- the **newton** (N) is the unit of force
- the **newton per kilogram** (N/kg) is the unit of gravitational field strength
- the **nanometre** (nm) is the unit of wavelength
- the **light year** (ly) is a large unit of length used in astrophysics.

Section 8b Motion in the universe

The universe

❑ The **universe** is a vast space which is continually expanding. There are regions within this space known as **galaxies** where billions of stars are grouped together.
(One billion = one thousand million, 1 000 000 000 or 10^9.)

❑ Each brightly coloured region in the diagram below represents a galaxy containing billions of stars.

8b Motion in the universe

☐ There are billions of galaxies distributed throughout the universe.

☐ Distances in space are commonly described in units known as a **light year (ly)**. This is because the distances are very, very large. **One light year (1 ly) is the distance light will travel in one year**.

In one year there are approximately 365 days; each day is 24 hours long, each hour has 60 minutes and each minute has 60 seconds. Therefore one year is:

365 × 24 × 60 × 60 = 31 536 000 s

Light travels at 3×10^8 m/s (300 000 000 metres in one second)

$s = v \times t$ = 300 000 000 × 31 536 000
= 9 460 800 000 000 000 m = 9.46×10^{15} m

One light year is approximately 9.46 million million kilometres.

☐ Each galaxy spans thousands of light years. The distance **between** galaxies is even bigger. For example, the galaxy we live in, known as the Milky Way, spans about 100 000 light years, but one of the nearest, Andromeda, is 2.5 million light years away.

☐ If one light year in space was equivalent to one kilometre, you would travel around the Earth approximately 60 times to travel the equivalent distance from the Milky Way to the Andromeda galaxy.

☐ A car travelling at 20 m/s would take approximately 65 million years to reach our nearest star, Proxima Centauri, which is 4.2 ly away.

- ❏ The diagram below shows our Milky Way galaxy and where our Sun and solar system are located in relation to the rest of the galaxy.

925 000 000 000 000 000 km (9.25 × 10^{20} m)

Our solar system

- ❏ Our Sun is a star in a spiral galaxy with 'arms' of stars, called the **Milky Way**. The Sun is about halfway out from the centre of the galaxy, on one of the arms as shown above.

- ❏ Our solar system is only a very small part of the Milky Way galaxy. If our Sun was the size of a grain of sand, the Milky Way would stretch across the continent of North America.

- ❏ The Sun is a star at the centre of our solar system. It appears much bigger than other stars in the sky because it's much closer to Earth. In terms of stars, it's rather average; there are much bigger stars than our Sun out there. For example, Alpha Scorpii A is 690 million times the size of our Sun.

- ❏ Our solar system is made up of planets, dwarf planets, asteroids and comets, which all **orbit** the Sun. The Earth is one of eight planets in our solar system. Smaller dwarf planets include Eris, Ceres and Pluto.

- ❏ Planets are not hot enough to emit their own light. We only see them because they are illuminated by (reflect light from) the Sun.

☐ The diagram below shows all the planets and their relative sizes. It does not show their relative distances from the Sun. To show their relative sizes and distances on the same diagram is impossible, because the distances are so large.

☐ The planets that make up our solar system in order from closest to the Sun are given below.
- Mercury is the closest planet to the Sun.
- Venus is almost the same size as the Earth.
- Earth is the only planet in the solar system known to support life. It contains free oxygen and oceans of liquid water.
- Mars, known as the red planet, has a thin atmosphere made mainly of carbon dioxide.
- Jupiter is bigger than all the planets put together. The surface isn't solid; it is a planet made mainly from gas.
- Saturn is also mainly a gas giant. It is surrounded by rings of ice and ice-coated rocks. These range in size from grains to boulders.
- Uranus is another gassy planet. It also has rings but they are much fainter than those of Saturn.
- Neptune is very similar in size to Uranus and it is mostly made up of ice and gas.

Gravitational force

☐ A galaxy containing billions of stars is held together by the **force due to gravity**, or **gravitational force**.

☐ The force due to gravity is the force of **attraction** between any two objects with mass. For example, we are attracted to the Earth and we have seen in section 1 (page 25) that the force due to gravity acting on a mass on Earth is called **weight**.

$W = m \times g$ where g is the **gravitational field strength**.

❏ The size of the gravitational field strength depends on:
- the size of the masses involved; the greater the mass of either, the greater the gravitational force of attraction
- the distance between the objects; doubling the distance between the objects reduces the gravitational force by one quarter.

❏ The gravitational field strength near to a planet varies depending on the mass and size of the planet. For example, *g* on Earth is 10 N/kg but on the Moon it is only 1.6 N/kg. Close to the Sun the gravitational field strength is 270 N/kg. This means that a man with a mass of 70 kg would weigh 700 N on Earth, 112 N on the Moon and 18900 N close to the Sun (if he didn't vaporise first). Other examples are given in the table below.

Planet	*g* in N/kg
Venus	8.8
Mars	3.8
Saturn	10.4
Jupiter	25

Objects in orbit

❏ An attractive force is required to keep a planet in orbit around the Sun, and a moon around a planet. This force is the gravitational force that exists between all masses. A planet, moon, asteroid, comet or any other object would travel in a straight line if no force were acting on it (see pages 20-21).

❏ In our solar system the Sun's gravitational field keeps the planets, dwarf planets, comets and asteroids in orbit round the Sun.

❏ Most of the planets have natural satellites, called moons, that orbit them. Some planets, such as Jupiter, have many moons. The gravitational force on each moon from its planet keeps it in orbit.

- ❏ Objects in orbit have both gravitational potential energy **GPE** and kinetic energy **KE**. **GPE** is the energy an orbiting object has because of its position in the gravitational field; for example, the Earth has **GPE** because of its position in the Sun's gravitational field. The Earth also has **KE** because it is moving.

- ❏ Energy can be transformed from **GPE** to **KE** and vice versa but the total energy never changes (see page 123 and pages 138-43).

- ❏ As the distance from the Sun increases, the time for a planet to complete an orbit increases because:
    - the orbital distance increases
    - the speed of a planet in its orbit decreases as gravitational force decreases.

Planets

- ❏ The planets are kept in orbit by the gravitational pull of the Sun. They mostly orbit around the Sun in a **near circular path**, much like electrons orbiting the nucleus of an atom. Planets in circular orbits would remain the same distance from the Sun (constant **GPE**) and travel at constant speed (constant **KE**).

- ❏ Different planets take different amounts of time to go around the Sun. A single orbit is called the planet's year, and the further out a planet is, the longer its year, because the orbital circumference is larger and the planet is also travelling slower.

- ❏ Astronomers use telescopes to help them work out the sizes of the planets, moons and the Sun. They can also work out their orbital paths and the distances between them.

- ❏ The planets all travel around the Sun in the same direction and in almost the same plane. The following diagram gives a sense of what these orbits look like (it is not to scale).

8 Astrophysics

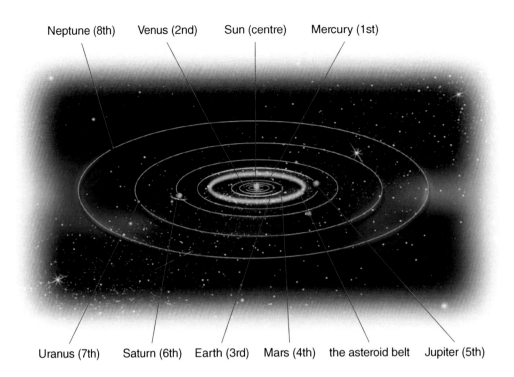

Neptune (8th) Venus (2nd) Sun (centre) Mercury (1st)

Uranus (7th) Saturn (6th) Earth (3rd) Mars (4th) the asteroid belt Jupiter (5th)

Dwarf planets

- Pluto was demoted in 2006 from a planet to a new classification known as a dwarf planet. It is a small icy body. Its elliptical (squashed circle) orbit takes almost 250 Earth years. Ceres and Eris are other examples of dwarf planets.

Asteroids

- Asteroids are made up of metal and rock and are much smaller than planets, usually with irregular shapes. Some larger asteroids are almost spherical and are now known as dwarf planets. The asteroids have diameters ranging from approximately 3 km to 1000 km.

- Most of them are found in an 'asteroid belt', in orbit around the Sun between the planets Mars and Jupiter. The dwarf planet Ceres is found here.

- The gravitational field of the Sun causes asteroids to orbit. Some of them have elliptical (squashed circle) orbits.

- Asteroids can crash into each other. This can cause them to break apart or their orbit to change.

Comets

- Comets are lumps of ice and dust, with 'heads' often several kilometres across, which orbit around the Sun.

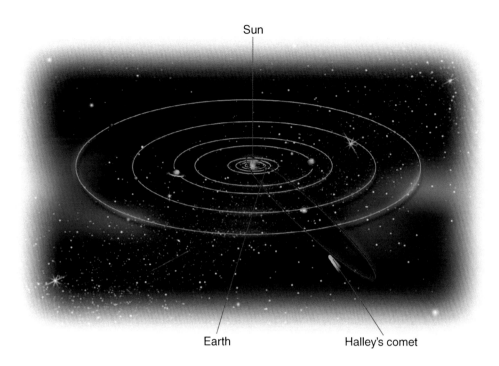

- Comets have **highly elliptical** (squashed circle) orbits, which take them from close to the Sun, to the edge of our solar system. When the comet is close to the Sun it travels very fast because the gravitational force of attraction is high. It slows as it moves away when the gravitational force becomes less.
  Remember: The total energy of a body in orbit is the sum of the *GPE* and the *KE*. As the comet approaches the Sun its *GPE* falls and it gains *KE*.

- The time period of a comet's orbit is a constant, which may vary for each comet from a few years to many thousands of years. For example, for many years Halley's comet has been visible from the Earth every 75 or 76 years.

- When a comet passes close to the Sun, the heat from the Sun causes the particles of dust and gas to stream off the icy 'head' of the comet into space, forming a huge 'tail', which is visible as thecomet passes the Earth, because it reflects the sunlight. Each time the comet passes close to the Sun, it loses some of its mass. Comets leave a trail of debris behind them.

Satellites

- A satellite is an object such as a moon or a machine that orbits a larger object in space. For example, our Moon orbits the Earth.

- Satellites are kept in orbit by gravitational force. There are two types of satellite.

Natural satellites

- A natural satellite is any celestial body outside the Earth's atmosphere (for example, the Moon, planets, asteroids, comets) that orbits a larger body. Examples are the Moon orbiting the Earth or the Earth orbiting the Sun.

- Many planets have moons that orbit around them; many stars are known to have planets orbiting them.

- The Moon is influenced by the Earth's gravitational field and circles our planet once every 27.3 days.

- The Earth is influenced by the Sun's gravitational field and circles the star once every 365.25 days.

Artificial satellites

- Artificial satellites are machines that have been placed into orbit by humans. Artificial satellites orbiting the Earth have several uses including:
  - telecommunications (transmitting information between distant parts of the Earth)
  - weather forecasts
  - observing the Earth's surface
  - space telescopes gathering information from space
  - satellite navigation systems ('satnav')
  - scientific experiments, in the case of the manned International Space Station.

- Satellites in lower orbits travel faster than those in higher orbits. Satellites in higher orbits take longer to make one orbit, i.e. they have a longer time period.

Orbital speed

❑ The speed at which an object orbits is known as its **orbital speed**.

❑ The further the object is from the body that it orbits:
- the longer it takes to complete an orbit
- the slower it moves.

❑ For example, the closest planet to the Sun, Mercury, takes 88 Earth days to complete an orbit. But Neptune, the furthest planet from the Sun, takes 60 190 Earth days (approximately 165 Earth years) to complete an orbit.

❑ The graph below shows the distances, in millions of kilometres, of the planets from the Sun and the speed at which they orbit around it. The curve shows that the orbital speed decreases at a decreasing rate with distance.

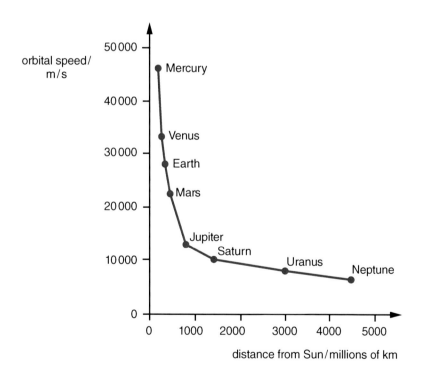

8 Astrophysics

❑ The orbital speed, orbital radius and time period are related by the following equation:

$$\text{orbital speed} = \frac{2 \times \pi \times \text{orbital radius}}{\text{time period}}$$

$$v = \frac{2 \times \pi \times r}{T}$$

v = orbital speed (m/s)
r = orbital radius (m)
T = time period (s)

N.B. The orbital radius is the distance between the centres of the bodies.

Example 8b (i)

The Earth is 150 million kilometres from the Sun, and takes one year to complete its orbit.
Calculate the orbital speed of the Earth in m/s. The equation linking orbital speed, time period and orbital radius is:

$$v = \frac{2 \times \pi \times r}{T}$$

Answer

Step 1 List all the information in symbol form and change into appropriate and consistent SI units if required.

r = 150 million kilometres = 150 000 000 000
  = 1.5×10^{11} m
T = 1 year = 365 × 24 × 60 × 60 = 31 536 000 s
v = ?

Step 2 Calculate the answer by putting the numbers into the equation.

$$v = \frac{2 \times \pi \times 1.5 \times 10^{11}}{31536000} = 29885.77 = 30000 \,\text{m/s}$$
(to 2 sig. figs)

ALWAYS REMEMBER TO STATE THE UNIT FOR CALCULATED QUANTITIES.

Example 8b (ii)

The Moon takes 27.3 days to orbit the Earth and its orbital speed is 1010 m/s. Calculate the radius of its orbit. The equation linking orbital speed, time period and orbital radius is:

$$v = \frac{2 \times \pi \times r}{T}$$

Answer

Step 1 List all the information in symbol form and change into appropriate and consistent SI units if required.

$v = 1010$ m/s
$T = 27.3$ days $= 27.3 \times 24 \times 60 \times 60 = 2358720$ s
$r = ?$

Step 2 Rearrange the equation.

$$v = \frac{2 \times \pi \times r}{T} \quad \Rightarrow \quad r = \frac{v \times T}{2 \times \pi}$$

Step 3 Calculate the answer by putting the numbers into the equation.

$$r = \frac{1010 \times 2358720}{2 \times \pi} = 379155966.8 = 3.8 \times 10^8 \text{ m}$$
(to 2 sig. figs)

ALWAYS REMEMBER TO STATE THE UNIT FOR CALCULATED QUANTITIES.

Example 8b (iii)

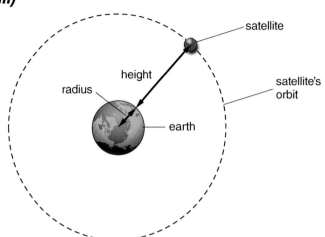

A geostationary satellite stays at exactly the same point above the equator, taking 24 hours to complete one orbit, just as the Earth takes 24 hours to rotate on its axis. The height of its orbit above sea level is 35800 km, and the radius of the Earth is 6370 km.
The equation linking orbital speed with time period and orbital radius is:

$$v = \frac{2 \times \pi \times r}{T}$$

(i) Calculate the radius of the orbit.
(ii) Calculate the orbital speed of the geostationary satellite.

Answer

(i) orbital radius = radius of Earth + height of satellite

$$= 6370 + 35800 = 42170 = 42200 \text{ km}$$
(to 3 sig. figs)

(ii) **Step 1** List all the information in symbol form and change into appropriate and consistent SI units if required.

$$r = 42200 \text{ km} = 42.2 \times 10^6 \text{ m}$$
$$T = 24 \text{ hours} = 24 \times 60 \times 60 \text{ s} = 86400 \text{ s}$$
$$v = ?$$

Step 2 Calculate the answer by putting the numbers into the equation.

$$v = \frac{2 \times \pi \times 42.2 \times 10^6}{86400} = 3068.87 = 3100 \text{ m/s}$$
(to 2 sig. figs)

ALWAYS REMEMBER TO STATE THE UNIT FOR CALCULATED QUANTITIES.

Section 8c Stellar evolution

- Stellar evolution is the process by which a star changes during its lifetime. On human timescales, a star doesn't change much at all. Stars are born, age and die over millions and billions of years. The lifetime of a star depends upon its **mass**.

Classification of stars by colour

- When we first look at stars they all appear to be white, but if we look more closely we notice a range of **colours**. Stars can be classified according to their colour. Their colour has a direct correlation with their surface temperature, so, for instance, a red star is cooler than a white star and a white star is cooler than a blue star. By looking at the colour of a star, astronomers can have an understanding of its temperature.

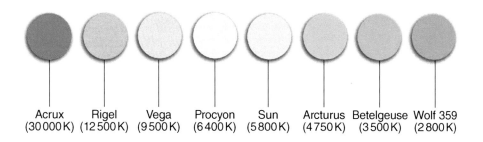

Acrux (30 000 K) Rigel (12 500 K) Vega (9 500 K) Procyon (6 400 K) Sun (5 800 K) Arcturus (4 750 K) Betelgeuse (3 500 K) Wolf 359 (2 800 K)

- Stars are classified with the letters O, B, A, F, G, K and M, where O stars are blue stars with the highest temperature and M stars are red stars with the lowest temperature.

- Our Sun is a yellowish-white star in class G and its surface temperature is about 6000 K.

The life cycle of a star

Nebula to protostar

- Stars begin in a cloud of dust and gas (mostly hydrogen) known as a **nebula**, most of which was left when previous stars blew apart in **supernovae**. The denser clumps of the cloud contract very slowly under the **force due to gravity**.

- As the cloud collapses, it fragments into smaller, denser regions, which themselves contract further to form stellar cores. These stellar cores, known as **protostars**, continue to contract, and heat up as they do so.

Main sequence stars

- ❏ The temperature of the core of the contracting protostar increases to the point where nuclear reactions begin (a few million degrees). At this point, **hydrogen** (H) nuclei fuse together to form **helium** (He) in the core of the star (see page 235).

- ❏ This process releases extremely large amounts of energy and creates enough pressure to stop the collapse of the star due to its **gravitational field**. The outward gas pressure from the nuclear fusion of hydrogen in the core balances the gravitational force trying to compress the star. This stage is called '**hydrogen core burning**' and the star emits huge amounts of radiation from its surface – it shines. It is now in a stable part of its lifetime, called the **main sequence**.

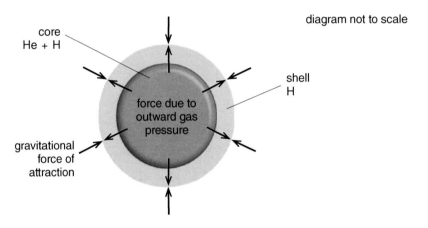

- ❏ For about 90% of its life, the star will continue to 'burn' hydrogen into helium and will remain a main sequence star. An average mass star – with a mass similar to the mass of our Sun – will remain in the main sequence for about 10 billion years.

> **Note**
>
> This 'burning' should not be confused with the chemical combustion of hydrogen in an atmosphere containing oxygen. There is no oxygen present.

Red giant stars

- ❏ Eventually the hydrogen in the core runs out, nuclear fusion stops and the core begins to contract again as the outward pressure stops but the gravitational force of attraction remains. The inert helium core continues to contract and heat up under the gravitational field of the star.

- ❏ The temperature of the helium core eventually becomes high enough to cause the hydrogen which surrounds the core to fuse. This is known as '**hydrogen shell burning**'.

- ❏ The core continues to contract and heat up until it is hot enough for the helium to fuse into carbon and oxygen. This is known as '**helium core burning**'. The fusion of helium releases enormous amounts of energy.

- ❏ This release of energy pushes the outer layers of the star even further out and causes the star to expand rapidly and become much larger and brighter. The outer layer is cooler than the inner core giving it a reddish colour. The star is now a **red giant**, a star with a mass similar to the Sun's, but much larger in size.

White dwarf stars

- ❏ When the helium in the core runs out, nuclear fusion stops. The core made of carbon and oxygen begins to contract again as the outward pressure stops but the gravitational force of attraction remains. The inert carbon–oxygen core continues to contract and heat up under the gravitational field of the star.

- ❏ The temperature of the core eventually becomes big enough to cause the helium that surrounds the core to fuse. This is known as '**helium shell burning**'.

- ❏ The core continues to contract and get hotter. This time the temperature is not sufficient to cause any further fusion in the core.

- ❏ The core continues to contract under its own weight but it does not have enough weight to collapse completely. The situation becomes unstable, as there are now two outer burning shells, one made of helium and one made of hydrogen.

- ❏ This instability causes the star to eject its outer layer, and lose mass.

- ❏ The core of the star is now a very dense hot solid known as a **white dwarf**, a star with a mass similar to the Sun's, but much smaller in size. This will eventually cool to a 'black dwarf'.

Red supergiant stars

- Stars with higher mass evolve in a similar way up to the main sequence stage. A massive star only remains in this stage for millions rather than billions of years.

- When the hydrogen in the core runs out, the star swells and becomes a **red supergiant**, a star with a mass higher than the Sun's, and very much larger in size.

- Unlike lower-mass stars, the contracting carbon–oxygen core of high-mass stars can heat up enough to cause fusion, this time forming neon.

- This process of core fusion followed by core contraction and shell fusion is repeated in a series of nuclear reactions. These produce successively heavier elements, until iron is formed in the core.

- Iron cannot be fused into heavier elements. The star has therefore run out of fuel and collapses under its own gravity.

- What happens next depends on the mass of the star. If it's about three to ten times the size of our Sun, it will explode in a **supernova** and leave behind a **neutron star**.

- If the mass of the star is even bigger, such as in very large giants or supergiants, a **black hole** is formed. A black hole is a point in space with an extremely large gravitational field.

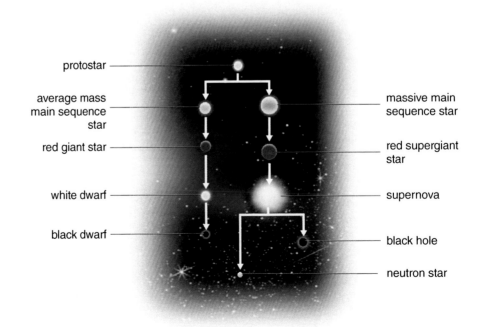

Brightness of a star

○ When we look at the sky at night, we see stars of varying brightness, but the brightness is affected by how far away the star is from us. Stars that are faint but close to Earth could appear brighter to us than brighter stars that are further away. For example, from Earth the planet Venus appears brighter than any other star in the sky. In fact, Venus is less bright than the stars; it is just closer to us.

○ To compare the actual, or absolute, brightness of stars, we should therefore state what it would be at a standard distance. Astronomers use a **standard distance** of 32.6 light years. Comparing stars in this way gives us an **absolute magnitude** of brightness.

Absolute magnitudes for familiar stars		
Name	Spectral class	Absolute magnitude
Sun	G	+4.8
Procyon	F	+2.7
Vega	A	+0.6
Acrux	B	−4.6
Betelgeuse	M	−5.1

N.B. The lower the value of absolute magnitude, the greater the brightness.

○ The Sun is the brightest object in our sky, so its apparent magnitude of brightness is very large. But when we compare it to other stars on the absolute magnitude scale, we see that there are other much brighter stars, for example, Betelgeuse is the brightest star in the table above.

The Hertzsprung–Russell (HR) diagram

○ In the early twentieth century, two scientists, Ejnar Hertzsprung and Henry Norris Russell, independently produced a scatter diagram comparing the absolute magnitude of a star and its colour (i.e. its surface temperature, see page 249), showing that the relationship was not random but that the stars fell into distinct groups.

○ The **Hertzsprung–Russell diagram** (**HR diagram**) is shown below.
N.B. The temperature axis is marked the opposite way to the normal convention – temperature decreases from left to right. While the numbers for absolute magnitude also decrease from bottom to top, remember that the lower the value of absolute magnitude, the greater the brightness.

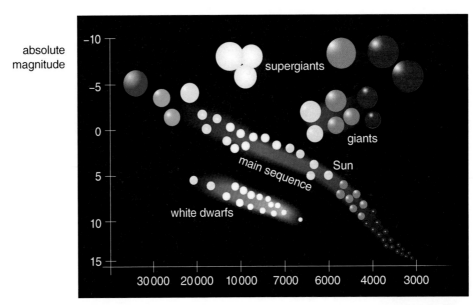

○ All main sequence stars lie on or very near a diagonal line on the diagram.

○ As a star evolves, its position in the HR diagram changes, following a characteristic path that depends on what chemicals the star is made of and its mass. In other words, its position in the above diagram will change over time.

Section 8d   Cosmology

- We do not know exactly how the universe was formed nor how big it actually is. We do know that gravitational forces have had a significant impact on what has happened. Gravitational forces will continue to influence the future of our universe.

The Big Bang

- The **Big Bang theory** is one of the most popular theories on how the universe was formed. According to this theory the universe began its existence 13.7 billion years ago. All matter and energy in the universe came from an extremely hot and dense point known as a 'singularity'. Nothing existed before this singularity – everything that we know today came after this event.

- It is believed that the universe suddenly expanded from this extremely hot, extremely dense and extremely small singularity, then continued expanding and cooling.

- Cosmologists believe this to be the order of events:
    - In a tiny fraction (10^{-36}) of a second, the universe grew from this singularity to become bigger than a galaxy during what is called the inflationary period. At that time the universe was empty of matter but contained large amounts of dark energy, the energy scientists believe is responsible for the force that keeps our universe expanding.
    - Energy and matter were inseparable. Then energy and matter became separate and tiny elementary particles of matter and anti-matter were formed.
    - In this first second, interactions between these tiny elementary particles meant that most matter and anti-matter annihilated each other, but some matter survived and produced larger, more stable sub-atomic particles called protons and neutrons.
    - Over the next three minutes, the temperature fell to below 1 billion degrees Celsius. It was now cool enough for the protons and neutrons to come together, forming hydrogen and helium nuclei.
    - After 300 000 years, the Universe had cooled further to about 3000°C and atomic nuclei could capture electrons to form simple atoms of hydrogen and helium.

- After about 500 million years the first stars and galaxies evolved and the expansion and cooling continued to become the universe as it is today, filled with billions of stars, countless planets, moons, asteroids, gas clouds and vast amounts of empty space.
- The universe continues to cool.

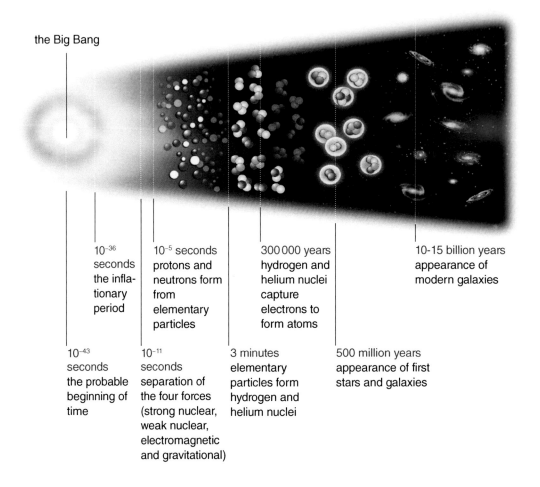

- 10^{-36} seconds the inflationary period
- 10^{-5} seconds protons and neutrons form from elementary particles
- 300 000 years hydrogen and helium nuclei capture electrons to form atoms
- 10-15 billion years appearance of modern galaxies
- 10^{-43} seconds the probable beginning of time
- 10^{-11} seconds separation of the four forces (strong nuclear, weak nuclear, electromagnetic and gravitational)
- 3 minutes elementary particles form hydrogen and helium nuclei
- 500 million years appearance of first stars and galaxies

○ Evidence that supports the Big Bang theory is:
- the observation of **red-shift** in the spectra of distant galaxies (see pages 257-58) shows that galaxies are moving away from us at speeds proportional to their distance, suggesting the universe was once compact and started from a single point
- the discovery of **cosmic microwave background** (**CMB**) radiation, which is uniform across the sky and does not appear to come from any one star or galaxy. This suggests that it is the left-over radiation from an extremely hot original source that has cooled over time. The source currently has a temperature of around 3K, just as predicted by the Big Bang theory.

Doppler red-shift

- As discussed in section 3 (see pages 115-16), the Doppler effect is the apparent change in frequency of a wave caused by the relative motion between the source of the wave and the observer.

- If a distant star and the Earth were stationary objects, the frequency of light observed on the Earth from the star would appear the same as the frequency from a similar source on Earth.

- Visible light is dispersed to form a spectrum of colours (see page 112). When astronomers view the light from a star or a galaxy, they see black lines in the spectrum. These black lines are caused by various elements absorbing particular wavelengths of light; for example, our Sun contains helium and hydrogen and the (simplified) spectrum from the Sun is shown below. This is known as an **absorption spectrum**.

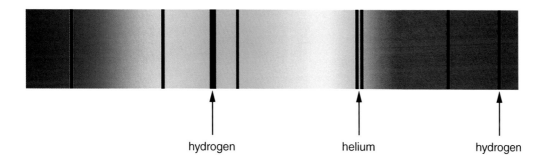

hydrogen   helium   hydrogen

- If a star or galaxy containing hydrogen and helium were moving **away** from the Earth, the frequency of these black lines would appear **lower** (longer wavelength). This is known as **red-shift**. The visible light observed is shifting towards the red end of the spectrum.

- If a star or galaxy containing hydrogen and helium were moving **towards** the Earth, the frequency of these black lines would appear **higher** (shorter wavelength). This is known as **blue-shift**. The visible light observed is shifting towards the blue end of the spectrum.

○ The diagram below gives an example of an absorption spectrum and shows how the black lines shift to the red end of the spectrum if the star is moving away from an observer, or to the blue end if it is moving towards the observer.

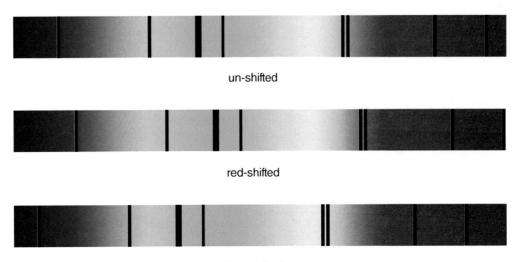

un-shifted

red-shifted

blue-shifted

○ When we view stars in a nearby rotating spiral galaxy, some stars will be moving towards us and others will be moving away from us as the galaxy rotates. We observe both blue- and red-shift.

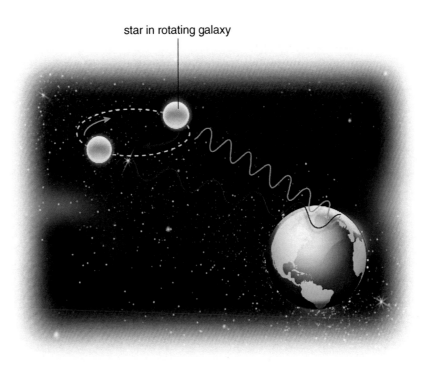

star in rotating galaxy

8d Cosmology

○ This shift in frequency (and wavelength) relates to the speed of the star moving towards or away from the Earth. The bigger the shift, the faster the star is moving.

○ When observing very distant galaxies, the light from individual stars cannot be distinguished; the light from the whole galaxy is analysed. The Doppler shift tells us whether the galaxy is moving towards or away from us, and how fast.

○ The change in wavelength, reference (original) wavelength, velocity of a galaxy and the speed of light are related by the equation:

$$\frac{\text{change in wavelength}}{\text{reference wavelength}} = \frac{\text{velocity of a galaxy}}{\text{speed of light}}$$

$$\frac{(\lambda - \lambda_0)}{\lambda_0} = \frac{\Delta\lambda}{\lambda_0} = \frac{v}{c}$$

v = velocity of galaxy (m/s)
c = speed of light in a vacuum (m/s)
λ = shifted wavelength (m)
λ_0 = reference (original) wavelength (m)

N.B. The wavelength of light is very small and is measured in **nanometres** (nm). A nanometre is 0.000000001 m or 10^{-9} m.

Example 8d (i)

An astronomer views the absorption spectrum from a galaxy and observes that the blue-green hydrogen line is red-shifted towards the red end of the visible spectrum at a measured wavelength of 500 nm. Calculate the velocity of the galaxy.

The original wavelength of the blue-green hydrogen line is 486 nm and the speed of light in a vacuum is 3×10^8 m/s.

The equation linking the change in wavelength to the speed of light and the velocity of the galaxy is:

$$\frac{(\lambda - \lambda_0)}{\lambda_0} = \frac{\Delta\lambda}{\lambda_0} = \frac{v}{c}$$

Answer

Step 1 List all the information in symbol form and change into appropriate and consistent SI units if required.

$\lambda = 500\,\text{nm}$
$\lambda_0 = 486\,\text{nm}$
$c = 300\,000\,000 = 3 \times 10^8\,\text{m/s}$
$v = ?$

We don't need to convert the wavelengths in nm to m because we're dealing with a ratio.

Step 2 Rearrange the equation.

$$\frac{(\lambda - \lambda_0)}{\lambda_0} = \frac{v}{c} \quad \Rightarrow \quad v = \frac{(\lambda - \lambda_0)}{\lambda_0} \times c$$

Step 3 Calculate the answer by putting the numbers into the equation.

$$v = \frac{(\lambda - \lambda_0)}{\lambda_0} \times c = \frac{(500 - 486)}{486} \times 3 \times 10^8$$

$v = 8641975 = 8.64 \times 10^6\,\text{m/s}$ (to 3 sig. figs)

ALWAYS REMEMBER TO STATE THE UNIT FOR CALCULATED QUANTITIES.

- Edwin Hubble was the first to observe red-shift in the spectra from Earth and distant galaxies; he also showed that this shift was bigger for galaxies further away from Earth. No distant galaxies observed showed blue shift.

8d Cosmology

○ Hubble's Law states that a galaxy's recessional velocity (its velocity away from us) is directly proportional to the galaxy's distance from Earth.

○ A graph of recessional velocity against distance from Earth is a straight line passing through the origin.

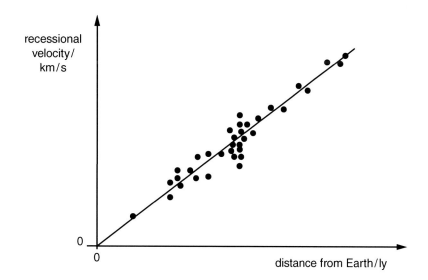

○ This tells us the **universe is expanding**, and that galaxies further away from the Earth are moving away from us faster, as predicted by the **Big Bang theory**.

Note

Remember: Red-shift means that the object, such as a star emitting light waves, is moving further away from the observer, such as an astronomer on Earth. The bigger the change in frequency towards the red end of the visible spectrum, the faster the star is moving.

Additional support material

1 Physical quantities and units

Separate Science in grey tint

Quantity	Symbol	Unit (usual unit **bold**)
Forces and motion		
length	l, h ...	**m**, mm, cm, km
distance	s, d	**m**, cm, km
time	t	**s**, min, h, ms
speed, velocity	u, v	**m/s**, km/h, cm/s
acceleration	a	**m/s²**
mass	m	**kg**, g, mg
weight	W	**N**
acceleration of free fall	g	**m/s²**
gravitational field strength	g	**N/kg**
force	F	**N**
momentum	p	**kg m/s**
moment of a turning force	M	**Nm**
Electricity		
charge	Q	**C**
current	I	**A**, mA
voltage/potential difference	V	**V**, mV
time	t	**s**
resistance	R	**Ω**
energy	E	**J**, kJ, MJ
power	P	**W**, kW, MW
Waves		
frequency	f	**Hz**, kHz
wavelength	λ	**m**, cm
angle of incidence	i	**degree** (°)
angle of reflection	r	**degree** (°)
angle of refraction	r	**degree** (°)

■ critical angle	*c*	**degree (°)**
■ refractive index	*n*	no unit
■ distance	*s*	**m**, cm, km
■ time	*t*	**s**, min, h, ms
■ speed, velocity	*v*	**m/s**, km/h, cm/s

Energy resources and energy transfers

■ mass	*m*	**kg**, g, mg
■ distance	*s*	**m**, cm, km
■ time	*t*	**s**, min, h, ms
■ speed, velocity	*v*	**m/s**
■ acceleration	*a*	**m/s²**
■ energy	*E*	**J**, kJ, MJ
■ work done	*W*, *E*	**J**, kJ, MJ
■ kinetic energy	*KE*	**J**, kJ, MJ
■ gravitational potential energy	*GPE*	**J**, kJ, MJ
■ force	*F*	**N**
■ power	*P*	**W**, kW, MW

Solids, liquids and gases

■ density	*ρ*	**kg/m³**, g/cm³
■ length	*l, h* ...	**m**, mm, cm, km
■ area	*A*	**m²**, cm²
■ volume	*V*	**m³**, cm³
■ distance	*s*	**m**, cm, km
■ speed, velocity	*v*	**m/s**, km/h, cm/s
■ acceleration	*a*	**m/s²**
■ force	*F*	**N**
■ pressure	*p*	**Pa**
■ energy	*E*	**J**, kJ, MJ
■ power	*P*	**W**, kW, MW
■ time	*t*	**s**, min, h, ms
■ current	*I*	**A**, mA
■ voltage	*V*	**V**, mV
■ temperature	*T*	**°C**
■ absolute temperature	*T*	**K**
■ specific heat capacity	*c*	**J/(kg °C)**, J/(g °C)

Magnetism and electromagnetism

■ current	I	**A**, mA
■ voltage	V	**V**, mV
■ power	P	**W**, kW, MW

Radioactivity and particles

■ half-life		**s**, **min**, **h**, year
■ activity		**Bq**
■ length	$l, h \ldots$	**m**, mm, cm, km
■ time	t	**s**, min, h, ms

Astrophysics

■ mass	m	**kg**, g, mg
■ distance	s	**m**, cm, km, ly
■ speed, velocity	v	**m/s**
■ acceleration	a	**m/s²**
■ force	F	**N**
■ time	t	**s**
■ gravitational field strength	g	**N/kg**
■ wavelength	λ	**m**, nm

2 How to use equations effectively

❏ Read each examination question carefully and decide which equation is required.

❏ Follow the steps below for calculations.

Step 1 List all the information in symbol form and change into appropriate and consistent SI units if required.

For example:
$v = 10\,cm/s = 0.1\,m/s$
$s = 5.0\,m$
$t = ?$

Step 2 Rearrange the equation (if necessary) so that the subject you are trying to find is on its own on the left-hand side of the equation.

For example: $v = \dfrac{s}{t} \quad \Rightarrow \quad t = \dfrac{s}{v}$

Step 3 Calculate the answer by putting the numbers into the equation, **remembering to include units for calculated quantities in your final answer**. (You may get marks for your working on the paper; remember this is the only way you can communicate with an examiner.)

For example: $t = \dfrac{5.0}{0.1} = 5.0\,s$

3 Equations

❏ **Learn all the equations** presented on the following pages. You must know them all for the Physics (Separate Science), or only the ones marked with a square for Science (Double Award).

The triangles provide a learning aid. They are **not** an alternative way of writing the equation.

Forces and motion

- average speed = $\dfrac{\text{distance moved}}{\text{time taken}}$

  $v = \dfrac{s}{t}$

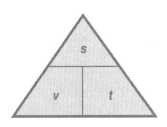

- acceleration = $\dfrac{\text{change in velocity}}{\text{time taken}}$

  $a = \dfrac{(v - u)}{t}$

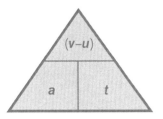

- force = mass × acceleration

  $F = m \times a$

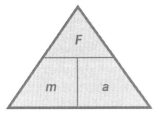

- weight = mass × gravitational field strength

  $W = m \times g$

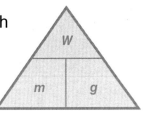

- momentum = mass × velocity

  $p = m \times v$

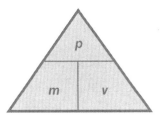

- moment = force × perpendicular distance from the pivot

  $M = F \times d$

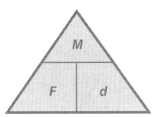

Electricity

- charge = current × time

  $Q = I \times t$

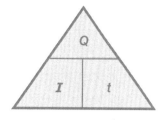

- voltage = current × resistance

  $V = I \times R$

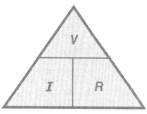

- energy transferred = charge × voltage

  $E = Q \times V$

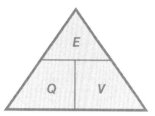

- electrical power = current × voltage

  $P = I \times V$

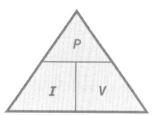

Waves

- wave speed = frequency × wavelength

  $v = f \times \lambda$

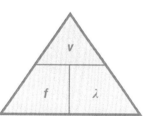

- refractive index = $\dfrac{\text{sine of the angle of incidence}}{\text{sine of the angle of refraction}}$

  $n = \dfrac{\sin i}{\sin r}$

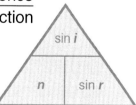

- sine of critical angle = $\dfrac{1}{\text{refractive index}}$

  $\sin c = \dfrac{1}{n}$

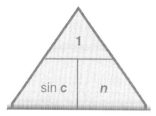

Energy resources and energy transfers

- work done = force × distance moved in the direction of the force

  $W = F \times d$

  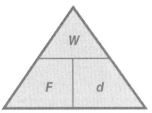

- energy transferred = work done

- kinetic energy = ½ × mass × speed²

  $KE = \dfrac{1}{2} \times m \times v^2$

  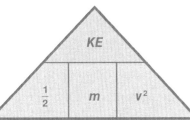

- gravitational potential energy = mass × gravitational field strength × height

  $GPE = m \times g \times h$

  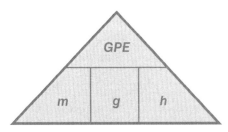

- efficiency = $\dfrac{\text{useful energy output}}{\text{total energy input}} \times 100\%$

- efficiency = $\dfrac{\text{useful power output}}{\text{total power input}} \times 100\%$

Solids, liquids and gases

- pressure = $\dfrac{\text{force}}{\text{area}}$

    $p = \dfrac{F}{A}$

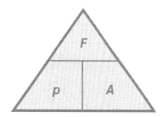

- $\dfrac{\text{pressure}}{\text{difference}}$ = height × density × $\dfrac{\text{gravitational}}{\text{field strength}}$

    $p = h \times \rho \times g$

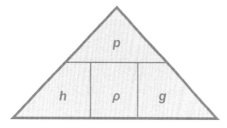

- density = $\dfrac{\text{mass}}{\text{volume}}$

    $\rho = \dfrac{m}{V}$

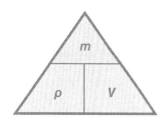

Magnetism and electromagnetism

- For a transformer:

    $\dfrac{\text{input (primary) voltage}}{\text{output (secondary) voltage}} = \dfrac{\text{primary turns}}{\text{secondary turns}}$

    $\dfrac{V_p}{V_s} = \dfrac{N_p}{N_s}$

- For a 100% efficient transformer:

    $\dfrac{\text{voltage in}}{\text{primary}} \times \dfrac{\text{current in}}{\text{primary}} = \dfrac{\text{voltage in}}{\text{secondary}} \times \dfrac{\text{current in}}{\text{secondary}}$

    $V_p \times I_p = V_s \times I_s$

Additional support material

4 Working with numbers

Understanding significant figures

- Numbers can be expressed in many different ways. Let us say the calculated value for a particular quantity was 2. One student may write it as 2 and another as 2.0. They may appear the same but they mean different things. 2 has one significant figure. Writing 2 means the value is 1.5 or above but below 2.5. 2.0 has two significant figures. Writing 2.0 means the value is 1.95 or above but below 2.05.

- In physics calculations, numbers may appear with many digits on your calculator, e.g. 5.046754327. Expressing your answer like this is **wrong** because it claims you know the answer far more precisely than any instrument you have used or any information or data you have been given. For example, let us say you are calculating the average time for an oscillation. If you use the answer above, you are claiming incorrectly that you can measure time with a stopwatch to this high degree of precision.

- As a general rule you should not give an answer to more significant figures than the least precise figure you have been given in the data. For instance, if you are asked to find the density of an object whose mass is 6.5 kg and whose volume is $0.72 \, m^3$, your calculation would be as follows:

$$\rho = \frac{m}{V} = \frac{6.5}{0.72} = 9.0277777 \, kg/m^3$$

The two values of data you are given are both to two significant figures and so your answer should also be given to two significant figures, in this case $9.0 \, kg/m^3$.

- Giving answers to **two or three significant figures is normally acceptable for the majority of Edexcel International GCSE Physics** questions. It is usually possible to express values from a graph to **three significant figures**.

Rounding to three significant figures

- Rounding means reducing the digits in a number while trying to keep its value similar.

- In the number 6.0469812 the fourth digit is above 5, so you would increase the third digit by one unit, e.g. 6.05. This is **rounding up**.
  In the number 6.0429812 the fourth digit is below 5, so you leave the third digit as it is, e.g. 6.04. This is **rounding down**.

- **Remember:** Round up if the digit to the right is 5 or above and round down if it is below 5.

Standard form

- Many very large or very small numbers in physics are expressed in standard form, also commonly known as **scientific notation**.

- Standard form consists of two parts, a number between 1 and 10 followed by × 10 to the power of a number, known as the **index**.

  e.g. 3.0×10^8 m/s is the speed of light

- The following numbers written out in full give:

  $1.0 \times 10^5 = 100\,000$
  $3.0 \times 10^8 = 300\,000\,000$
  $3.56 \times 10^7 = 35\,600\,000$

  The value of the index depends on how many places the decimal point has been moved. For example 35 600 000 becomes 3.56 by moving the decimal point 7 places to the left. The index is 7.

- To represent very small numbers, a negative index is used. The following numbers written out in full give:

  $2.1 \times 10^{-5} = 0.000021$
  $9.3 \times 10^{-8} = 0.000000093$
  $3.56 \times 10^{-4} = 0.000356$

  This time the decimal point moves to the right and the index is a negative number. For example 0.000356 becomes 3.56 by moving the decimal point 4 places to the right. The index is −4.

Additional support material

- Values can be entered in standard form on a calculator by using the EXP button. For example:

    7.6×10^4 = [7] [.] [6] [EXP] [4]

    3.2×10^{-3} = [3] [.] [2] [EXP] [(−)] [3]

    Do not include the 10. You would only use the 10 when using the [x^y] button, for example: [3] [.] [2] [×] [1] [0] [x^y] [(−)] [3] and so on.

    N.B. Some calculators have the label ×10^x instead of EXP. They both have the same function, so in the example above, EXP would be replaced with ×10^x.

- Sometimes, **prefixes** are used with SI units to simplify large or small values.

    kilo (k) = 10^3 milli (m) = 10^{-3}
    mega (M) = 10^6 micro (μ) = 10^{-6}
    giga (G) = 10^9 nano (n) = 10^{-9}
    tera (T) = 10^{12} pico (p) = 10^{-12}

    For example, you can write 2000 m as 2 km and 0.003 m as 3 mm.

    Always change prefixed units to base units but notice that the kilogram is the base SI unit for mass.

5   Graphs

You may be advised which set of numbers is to be plotted on the **x**-axis and the **y**-axis.

- *Choosing the appropriate scale for the x- or y-axis*
    - Find the maximum and minimum value needed for each axis. **Remember:** You do not have to start the axis from zero if the numbers given are all greater than zero unless you are trying to show that the two variables are proportional, in which case you must show that the line passes through the origin.
    - Make the scale easy to interpret. Choose a scale where each square on the graph paper equals 1, 2, 5, 10, 50 or 100. For smaller values use 0.01, 0.02, 0.05, 0.1 or 0.5.

- ❏ **The axes**
  - The **x**-axis will usually represent the **independent variable** and will rise in regular intervals, e.g. 2, 4, 6, etc. not 2, 4, 9, etc. The independent variable is the one you control. For example if you take a measurement every 10s, time is your independent variable.
  - The **y**-axis will usually represent the experimental results – the **dependent variable**. This will usually give rise to a straight line or a curve.
  - Label each axis with both variable and unit.

- ❏ **Drawing the graph**
  - The graph you draw (i.e. the points you plot) should cover as much of the graph paper as possible – three-quarters of the page is a good guide.
  - Plot crosses (x) or encircled dots (⊙) rather than dots (·) on your graph. Re-check any points that do not appear to fit the pattern (anomalous results).
  - Draw a smooth continuous line that will not necessarily pass through all the points, known as a **line of best fit**. If the graph looks like a straight line, then use a ruler.

- ❏ **Finding the gradient of a line**
  - Use $m = \dfrac{\Delta y}{\Delta x}$ to find the gradient of a line. Make sure you pick two points that are **on the line**. It is helpful to show the values you choose by drawing dotted lines from the axes (see below). Choose values as **far apart** as possible to give a more accurate gradient.

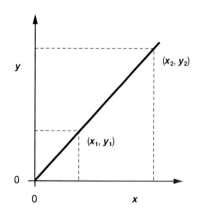

Example

Let (x_1, y_1) be (5.0, 50) and (x_2, y_2) be (10, 100).

$$m = \frac{(y_2 - y_1)}{(x_2 - x_1)} = \frac{(100 - 50)}{(10 - 5.0)} = 10$$

In many cases the gradient will have units. For example, if **y** is distance measured in metres and **x** is time measured in seconds, the gradient calculated above will be 10 m/s.

Understanding the graph you have plotted

❑ **Proportionality and linearity**

Many quantities in physics are **directly proportional** to each other. Many equations are derived from **straight-line (linear)** relationships.

The equation of a straight line on a graph is made up of a **y** term, an **x** term, and sometimes a number, and is written in the form of **y = (m × x) + c**.

❑ **Graph 1 – directly proportional**

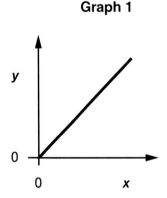

Graph 1

As **x** increases, **y** increases. The graph is a straight line and passes through the origin. Importantly the ratio of **x:y** is always the same. The graph formula is **y = (m × x)**, where **m** is the steepness of the line, also known as the gradient:

$$m = \frac{(y_2 - y_1)}{(x_2 - x_1)}$$

❑ *Graph 2 – straight line with y-intercept*

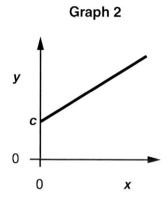

Graph 2

As **x** increases, so does **y** as before. The difference is when **x** = 0, as this time **y** is not zero. The graph formula is **y** = (**m** × **x**) + **c**, where **m** is the gradient. The **y**-intercept (the value of **y** when **x** = 0) is **c**.
N.B. This graph shows a linear relationship. It does not show a directly proportional relationship as the line does not pass through the origin.

❑ *Graph 3 – straight line with x-intercept*

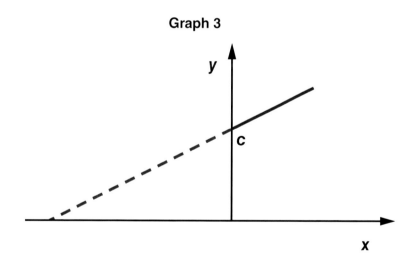

Graph 3

The graph equation is still **y** = (**m** × **x**) + **c**, and **c** is still the **y**-intercept.
Sometimes you will be required to extrapolate the graph, i.e. to extend the graph beyond the actual data. In graph 3 above, for instance, you could be asked to predict the value of **x** when **y** = 0.

Glossary

absolute magnitude the brightness of a celestial body if it were at a distance of 32.6 ly from an observer on the Earth

absolute zero the theoretical temperature at which all molecular motion ceases; equal to −273.15 °C

a.c. generator a device used to produce alternating current

acceleration how much an object's speed is increased per second – the rate of change of velocity of an object

acceleration of free fall the acceleration of an object falling freely under gravity

accurate how close a measured value is to a known value

action one of a pair of forces referred to in Newton's 3rd Law; the action causes an equal and opposite reaction

activity the rate of decay of nuclei in a radioactive sample

air resistance the frictional force on an object moving through air

alpha (α) particle a type of nuclear radiation consisting of a helium nucleus ejected from an unstable nucleus

alternating current (a.c.) electric current that changes its direction repeatedly and rapidly in a circuit; the charges flow one way and then the other way

ammeter an instrument for measuring electric current

amplified increased in size

amplitude the maximum height or disturbance of a wave from its central equilibrium (rest) position

angle of incidence the angle measured between the normal and a ray of light arriving at a surface

angle of reflection the angle measured between the normal and a ray of light reflecting at a surface

angle of refraction the angle measured between a refracted ray and the normal to a surface

anomalous odd results not in keeping with the rest of your results; not fitting the pattern

anti-matter matter composed of anti-particles which are elementary particles with the same mass but the opposite charge as the particles which make up matter; for example an anti-proton has the same mass as a proton but has a negative charge

atomic number see **proton number**

attraction a force that causes objects to move towards each other

average speed the speed calculated by dividing total distance moved by total time taken

background radiation the radiation in the surrounding environment that we are exposed to all the time

balanced equal in size but opposite in sign or direction and therefore adding to zero

beta (β⁻) particle a type of nuclear radiation consisting of a high – speed electron emitted from an unstable nucleus

Big Bang theory a scientific theory describing the origin of the universe

black hole a massive object with a gravitational field so strong that nothing can escape, not even light

blue-shift the change in wavelength of light, due to the Doppler effect, when stars in rotating galaxies move towards us

boiling point the temperature at which a liquid changes to a gas at normal pressure

Boyle's Law for a given mass of a gas at constant temperature, the volume of the gas is inversely proportional to the pressure

celestial of or relating to the sky or the universe

centre of gravity the point in an object where all the weight appears to be concentrated

chain reaction a reaction in which the reaction products promote further reactions

charge an electrical property, which can be positive or negative; a charged particle will experience a force if placed in an electric field

circuit breaker a safety device that switches off automatically when the current in it becomes too high; it can easily be reset

collision where two or more objects strike each other and each object exerts a force on the other(s)

comet a small icy body that orbits the Sun

component part of a mechanical or electrical system

compression a region of a sound wave in which the particles are pushed close together

conduction the process by which thermal energy is transferred, or by which charge flows

conductor a material that transfers thermal energy, or allows charges to pass through it easily

conserved maintained at a constant overall value; does not change; for example, energy into and out of a system

constriction the act of narrowing, such as the narrowing of a tube

consumer a person who purchases goods and services such as electrical energy for personal use

contamination unwanted pollution

control rods rods used in nuclear reactors to absorb the neutrons and control the output

convection the transfer of energy by hot liquid or gas rising and cold liquid or gas falling

conventional current the imagined flow of positive electric charge, from the positive terminal in a battery round the circuit to the negative terminal

coolant a fluid (liquid or gas) used to cool down a system

cosmic microwave background (CMB) radiation left over from the Big Bang

cosmology the study of the origin and evolution of the universe

count rate the number of decaying radioactive nuclei detected per second or minute

critical angle the angle of incidence above which total internal reflection occurs

current the rate of flow of electric charge in a circuit

decay see **radioactive decay**

deceleration (also known as negative acceleration) how much an object's speed is decreased per second; a negative rate of change of velocity of an object

density the mass per unit volume of a substance

dependent variable in an experiment, the variable being tested

depleted used up to the point of it running out, such as an empty petrol tank in a car

depth the distance from the top to the bottom, such as in a swimming pool

determine find a quantity as a result of calculation or from a graph; often used when the quantity cannot be measured directly

deviated moved away from the original intended course

diode a circuit component that allows current in one direction only

direct current (d.c.) electric current that is always in the same direction

directly proportional two quantities are directly proportional if their ratio is constant; as one increases, the other increases by the same percentage; if a graph of one quantity y is plotted against the other x, the graph is a straight line that passes through the origin (0, 0)

dispersion the splitting of white light into its component wavelengths (colours); for example, when white light falls onto a triangular prism

dissipate to disperse or scatter

domains groups of small atomic magnets that need to line up for a magnetic material to become magnetised

Doppler effect the change in frequency of a wave for an observer moving relative to the source of the wave

drag a type of friction (sometimes called air or water resistance) that opposes the motion of an object through a fluid

earthing when a charged object is connected to earth (or ground) and the charges flow to earth; for example, the casing of an electrical appliance is connected to the earth wire for safety

echo the reflection of sound from a surface, heard some time after the original sound

efficiency a fractional measure (usually expressed as a percentage) of how effectively energy or power is transferred into a useful form in comparison to the total energy or power

electric current see **current**

electric field a region in space in which an electric charge will experience a force

electromagnet a coil of wire with an iron core that becomes magnetic only when there is an electric current in the coil

electromagnetic induction a method of producing a voltage by moving a magnet relative to a coil of wire

electromagnetic (e.m.) spectrum the family of e.m. waves in order of their frequency, ranging from radio to gamma

electromagnetic (e.m.) waves energy travelling in the form of waves, which require no medium in which to travel

electron a very small subatomic particle that is negatively charged and exists in orbitals around the nucleus of an atom

electrostatic charges positive or negative charges that can be present on an insulator; insulators do not allow charges to move so they are static (stay still)

endoscope a fibre optic device used to image the inside of living bodies

equilibrium when there is no net moment and no net force acting on a body

evaporation the process by which a liquid changes to gas or vapour below its bo**exert** to make a physical effort such as when applying a force

extension the increase in the length of a spring when a load is attached

fission see **nuclear fission**

fluid a material that flows, such as any liquid or gas

fossil fuel a source of energy such as coal, oil and gas formed from the remains of dead plants and animals over millions of years

frayed cables electricity cables that are unravelled and have worn insulation

free fall an object falling under the influence of gravity alone

frequency the number of waves passing a point per second, or the number of vibrations per second

friction the force that opposes motion when two surfaces rub together

fuse a component designed to melt when a specified current value is exceeded, thus breaking the circuit

fusion see **nuclear fusion**

galaxy a system of millions or billions of stars

gamma (γ) ray highly penetrating electromagnetic radiation produced when an unstable atom decays

geothermal energy thermal energy produced in the Earth's core

gradient the ratio of the change in the quantity plotted on the y-axis to the corresponding change in the quantity plotted on the x-axis; here quantity plotted means, for example, distance/m rather than just distance

gravitational field strength the force in newtons exerted per kilogram of mass by gravity; at the Earth's surface this is approximately 10 N/kg

gravitational potential energy (*GPE*) the energy possessed by an object due to its relative position; for example, its height above the Earth's surface

half-life the average time taken for half the nuclei in a radioactive sample to decay and form a new element

hard magnetic material material that, once magnetised, retains its magnetism

Hertzsprung – Russell diagram see **HR diagram**

Hooke's Law the extension of a spring is proportional to the load producing it, provided that the limit of proportionality is not exceeded

HR diagram a scatter graph of stars showing the relationship between the stars' luminosities or absolute magnitudes and their temperatures or spectral classifications

hydroelectricity electricity produced using the gravitational potential energy of water stored in reservoirs in mountainous regions to turn turbines and drive generators

ideal gas a theoretical gas in which the molecules have negligible volume and do not interact with each other

image optical reproduction of an object using lenses or mirrors (see also **real image** and **virtual image**)

incident ray a ray of light arriving at or striking a surface

independent variable in an experiment, the variable you control

induce to give rise to or cause something to happen

induction see **electromagnetic induction**

infrared radiation the portion of the electromagnetic spectrum between microwaves and visible light that is sometimes known as thermal radiation

infrasound low-frequency sound, lower than 20 Hz (the normal limit of human hearing)

insulator a material that does not transfer thermal energy, or that does not or allow electrical charges to pass through it easily

interact to act in a manner that causes objects to have an effect on each other

interaction see **interact**

internal energy the total kinetic energy and potential energy of the particles within an object

inversely proportional as one quantity increases, the other quantity decreases so that their product is constant

ion an atom or molecule which has gained or lost electrons leaving it with a net negative or positive charge

ionisation the process by which a particle (atom or molecule) becomes electrically charged by losing or gaining electrons

ionising radiation charged particles or high-energy e.m. rays that ionise the material through which they travel

irradiation exposure to radiation

isotopes atoms of an element that have the same atomic number but different nucleon number (the same number of protons but a different number of neutrons)

kinetic energy (*KE*) the energy of an object due to its motion

lagging material providing thermal insulation

latent heat the energy needed to melt or boil a material without change of temperature

laterally inverted an image formed by a mirror which is reversed left to right compared with the object

light-dependent resistor (LDR) a resistor whose resistance varies according to the amount of light falling on it

light-emitting diode (LED) a diode that glows when current passes through it

light year the distance that light travels in one year, approximately equal to 9.46×10^{15} m

limit of proportionality the point beyond which the extension of a spring is no longer proportional to the load

linear in a straight line

load a force, often a weight

longitudinal wave a wave in which the particles of the medium through which the wave travels move backwards and forwards along the same line as the direction of travel

loudness how we perceive the amplitude of a sound wave

luminous gives out light

magnetic field a region in space around a magnet or electric current in which a magnet will feel a force

magnetic field lines imaginary lines that show the direction and strength of a magnetic field

magnetic material material that is attracted to or can be made into a magnet; steel and iron are examples of magnetic materials

magnified enlarged; for example, a magnified image is larger than the object

magnitude the size of a quantity; or the brightness of a star

mass the amount of matter in an object

mass number see **nucleon number**

matter material that has a mass and occupies space

mean value an average value

medium a material that waves can travel through, such as air, water or glass

melting point the temperature at which a solid melts to become a liquid at normal pressure

meniscus the curved surface of a liquid as seen, for example, within a measuring cylinder

microwaves electromagnetic waves with wavelengths from 1 m to 1 mm

model a mental or physical representation that cannot be observed directly; usually used as an aid to understanding

moderator a substance in a nuclear reactor that slows down neutrons

moment the turning effect of a force about a pivot; the product of force and perpendicular distance from the pivot

momentum the product of mass and velocity of a body; it is a measure of the quantity of motion in a body

moon a natural satellite of a planet

mutation a change in the genetic structure of a cell

nebula a cloud of gas and dust in outer space

negligible small enough to be ignored

neutral having no overall positive or negative charge

neutral having no overall positive or negative charge; also used where the net effect of forces is zero e.g. magnetic forces

neutron star a celestial object of very small radius and very high density

Newton's 1st Law an object will remain at rest or move at a steady speed in a straight line unless acted upon by an unbalanced force

Newton's 2nd Law an object will accelerate in the direction of an unbalanced force

Newton's 3rd Law when body A exerts a force on body B, body B exerts an equal and opposite force on body A

non-renewable energy from sources that will not be replenished in our lifetime, e.g. fossil fuels

normal a construction line drawn at right angles where a light ray strikes a surface, such as a mirror or glass block

nuclear energy the energy stored in the nucleus of an atom

nuclear fission the process by which energy is released by the splitting of a large heavy nucleus into two or more lighter nuclei

nuclear fusion the process by which energy is released by the joining of two small light nuclei to form a new heavier nucleus

nucleon a proton or a neutron, found in the nucleus of an atom

nucleon number (*A*) (also known as mass number) the number of protons and neutrons in the nucleus of an atom

nucleus the very small and dense centre of an atom made up of protons and neutrons

nuclide a 'species' of a nucleus characterised by its atomic and mass numbers

Ohm's Law the current in a metal conductor is directly proportional to the voltage across it provided the temperature remains the same

orbit the path of a satellite or planet around a celestial body or an electron around the nucleus of an atom

oscillation a repetitive motion, such as a pendulum swinging back and forth

oscilloscope an instrument for displaying and analysing the waveforms of electronic signals

parallax error the apparent change in position of a measurement caused by viewing the measurement at an angle less than or greater than 90 degrees to the scale

parallel circuit a circuit in which current has more than one path

penetrating passing through or into something

period see **time period**

periscope an optical device for viewing objects otherwise out of sight

perpendicular at right angles or 90 degrees

pitch how high or low a musical note sounds; is related to frequency: the higher the frequency, the higher the pitch

pivot the fixed point about which a lever turns

plane a flat two-dimensional surface

planet a celestial body that does not produce its own light and is illuminated by a star around which it orbits

potential energy energy that is stored and has the potential to do work

power the rate at which energy is transferred or work is done

precise how close repeated measurements of the same quantity are to each other; precise measurements do not necessarily indicate accuracy

pressure the force acting per unit area at right angles to the surface

Pressure Law for a fixed mass of an ideal gas at constant volume, the pressure is directly proportional to the temperature measured in kelvin

primary coil the input coil to a transformer

propagation the manner in which energy passes from one point to another

proportional when a change in one quantity is accompanied by a consistent change in the other

proton a positively charged particle found in the nucleus of an atom

proton number (Z) (also known as atomic number) the number of protons in the nucleus of an atom

protostar an early stage in the evolution of a star

radial moving or directed along the radius of a circle

radiation the transfer of energy by electromagnetic waves or by alpha or beta particles

radioactive decay the natural and random change of an unstable nucleus when it emits radiation

radio waves electromagnetic waves with wavelengths from thousands of metres to 30 cm

random a spontaneous event that cannot be predicted; for example, random decay means that it is not possible to predict when a particular nucleus will decay (but, statistically, it is possible to determine how many nuclei will decay in a given time)

range the difference between the minimum and maximum, or the maximum distance a particle can travel

rarefaction a region of a sound wave in which the particles are further apart

ray a narrow beam of light or other electromagnetic radiation

reaction one of a pair of forces referred to in Newton's 3rd Law; reaction is equal in size and opposite in direction to the action

reaction time the amount of time taken for a person to respond to a stimulus; for example the time between a car driver seeing a red light and applying the brakes

reading of an instrument; the information given by the scale or gauge of a meter or instrument

real image an optical image that can be formed on a screen

rectification the process of converting a.c. into d.c. with the use of one or more diodes

red giant a very large bright star in the final stages of evolution

red-shift the change in wavelength of light, due to the Doppler effect, from distant galaxies that are moving away from us

reflection the change in direction when a ray of light or sound wave bounces off a surface

refraction the change in direction of a light ray when passing from one material into another

refractive index the ratio of the sine of the angle of incidence to the sine of the angle of refraction; a material that has a large refractive index will refract light more than a material with a lower index

relay an electromagnetic switch

renewable energy energy from resources which are naturally replenished, e.g. solar, wind

repulsion a force that causes objects to separate and move in opposite directions

resistance a measure of how difficult it is for current to pass through a circuit or part of a circuit

resistor a component in an electrical circuit that resists current

resultant force the net force acting on a body when two or more forces are unbalanced; effectively the replacement of all forces acting on an object with one equivalent force

ripple a small, uniform wave on the surface of water

Sankey diagram a diagram summarising the energy transfers in a process

satellite an object in orbit around another object

scalar quantity a quantity with magnitude (size) only

secondary coil the output coil of a transformer

sensitivity response to change; a sensitive instrument gives a large reading for a small change in the quantity being measured

series circuit a circuit in which current has one path

short circuit a low-resistance connection between two points, causing a large current to flow

SI an internationally agreed system of units based on the metric system

significant figures the digits of a number that carry meaning

slip rings connectors that allow the passage of current to and from a coil in an a.c. generator

Snell's Law the ratio of sin *i* to sin *r* is a constant and is equal to the refractive index of the second medium (with respect to the first)

soft magnetic material material that, once magnetised, can easily be demagnetised

solar cell a device which converts energy from the Sun into electricity

solar system the region of the universe that contains our Sun, the Earth and Moon, seven other planets and their moons and satellites, as well as asteroids and comets

solenoid a long cylindrical coil of wire that becomes magnetised when a current is present in the coil

sound wave a longitudinal wave that carries energy away from the source of sound, as mechanical vibrations of a medium

specific heat capacity (s.h.c.) the energy needed to raise the temperature of one kilogram of a substance by one degree Celsius

spectrum colours of light separated out in the order of their wavelengths

speed the distance moved by an object per second

speed of light in a vacuum is 3×10^8 m/s

split rings connectors that allow the passage of current to and from a coil in a d.c. motor; reverse the direction of the current every half turn

star a luminous celestial body comprising of gas held together by gravitational force

static electricity electric charge held by a charged insulator

stationary standing still

stellar of or relating to the stars

sterilise to kill bacteria

stopping distance the distance a vehicle travels in coming to a stop; the sum of thinking distance and braking distance

streamlined smoothed and somewhat rounded in order to reduce air or water resistance

temperature a measure of how hot a body is, dependent on the average kinetic energy of its particles

terminal velocity the maximum velocity reached by an object falling through a fluid such as air; reached when the force due to its weight is equal to the force due to air resistance

thermal energy the internal energy a body has because of the motion of its particles

thermistor an electrical component whose resistance decreases with an increase in temperature and vice versa

time period the time taken for one complete cycle of oscillation, vibration, passage of a wave, or orbit

total internal reflection when all light is reflected back from a surface between materials; this happens as a result of the angle of incidence at the less dense material being greater than the critical angle

transducer any device or component that converts one form of energy into another

transformer a device used to change the voltage of an a.c. electricity supply

transmission lines power cables used to carry electricity from power stations to consumers

transverse wave a wave in which the vibrations or oscillations are at right angles to the direction of travel

turbine a machine similar to a fan with blades that rotate when air, steam or water passes through; used to generate electricity

ultrasound sound waves with frequencies greater than 20 000 Hz, which cannot be heard by the human ear

ultraviolet radiation the portion of the electromagnetic spectrum between visible light and x-rays, which can cause skin damage

uniform constant

vapour a substance in its gaseous state that can be liquefied by pressure alone i.e. without cooling

variable a letter or symbol that represents a number or value

variable resistor a component whose resistance can be manually altered

vector a quantity with both magnitude (size) and direction

velocity the speed of an object in a particular direction

vice versa the other way round

virtual image an image, such as that in a plane mirror, that cannot be formed on a screen

voltage a measure of the energy transferred per unit charge passing through a component; also a measure of the amount of energy transferred to electrical form per unit charge by an electrical power supply, like a battery

voltmeter a meter used for measuring the voltage between two points of a circuit

wavefront an imaginary surface joining all points of a wave affected in the same way; for example, the lines formed by the crests of ripples on a pond

wavelength the distance between two adjacent identical points on a wave

weight the downward force due to gravity acting on an object's mass

white dwarf a small very dense star

work done the product of force and distance moved in the direction of the force

x-rays a form of electromagnetic radiation with very short wavelengths

Glossary of examination terminology

This glossary (which is relevant only to science subjects) will prove helpful to candidates as a guide, but it is neither exhaustive nor definitive. The meaning of a term will depend, in part, on its context.

Add/Complete/Label: Requires the addition of information or labels to, for example, a graph, diagram or table of results.

Analyse: Examine the data in detail to provide an explanation.

Calculate: Obtain a numerical answer, showing relevant working.

Comment on: Consider the information given and make a judgement.

Deduce: Draw conclusions by making logical connections between pieces of information given.

Define/Give/State/Name (the term): Literally, give the meaning of the term; requires a concise answer, with little or no supporting evidence or argument.

Describe: Give an account in words (using diagrams if appropriate) of the main points of something, without giving reasons.

Describe and explain: Give an account, with reasons.

Design: Plan or invent a procedure.

Determine: Requires a quantitative answer; used about a quantity that cannot be measured directly but has to be obtained from a graph or calculation.

Discuss: Give a critical account by identifying the issues, exploring all aspects of the issues and investigating the issues by reasoning or argument.

Draw: Produce a diagram (with labels).

Estimate: Find, or roughly calculate, the approximate value of something from a graph or from a calculation.

Evaluate: Review information and form a conclusion, considering strengths, weaknesses, alternative actions and relevant data; come to a supported judgement of the quality of something.

Explain: Give reasons, and perhaps some reference to theory, as to why something happens; this can include mathematical explanations.

Find: A general term that may be interpreted as calculate, measure, determine, etc.

Give reasons: The only requirement is to give reasons.

Identify: Find some key point from information given in the question.

Justify: Give evidence to support some information given in the question or an earlier answer.

List: Requires a number of items/points, without further explanation.

Measure: The quantity concerned can be directly obtained from a suitable measuring instrument (e.g. time with a stopwatch, distance with a metre rule, etc.).

Plot: Produce a graph from information given by marking points accurately on a grid, and drawing a line of best fit through the points; the axes should be labelled with variables and units, and should be marked with a suitable scale.

Predict: Give an expected result, not by recall but by making a logical connection between other pieces of information.

Show that: Verify the statement given.

Sketch: Produce a freehand drawing or sketch a graph; for a graph, the axes should be labelled but they do not need scales, there should be a line or curve, and any important features should be noted (Does it pass through the origin? Does it have an intercept? etc.).

State what is meant by: Usually requires a definition, together with some comment about the context or significance of the terms.

Suggest: Use your knowledge to propose a solution to a problem in a novel context.

What, Why: Questions used to ask for information.

Index

absolute magnitude of stars 253
absolute temperature scale 180
absolute zero 180
absorption spectra 257–8
a.c. generators 210–11
acceleration 6, 8–9
   calculation from a velocity–time graph 10, 11
   and forces 21, 30–1
   Newton's Laws of Motion 22–3
   relationship to velocity and time 10, 16–18
   units of 2
acceleration of free fall 31, 138
accuracy 5
acid rain 152
aerosols 189–90
air resistance 29, 31–3, 140
airbags 39
alpha (α) particles 219, 227
   effects of magnetic fields and charged plates 223–4
   harmful effects 220
alpha decay 228
alternating current (a.c.) 54
   a.c. generators 210–11
ammeters 69
amperes (A) 62
amplifiers 82
amplitude of a wave 91
   sound waves 121, 122
angle of incidence 94, 102, 105
angle of reflection 94, 102
angle of refraction 105
artificial satellites 244
asteroids 242
atmospheric pressure 164
atomic number (proton number) 217
atomic structure 85, 217–18
average speed 7
   measurement of 15
   relationship to distance and time 8, 13

background radiation 220
balanced forces 20
bar magnets 196–7
becquerels (Bq) 217, 229
beta (β^-) decay 228
beta (β^-) particles 219
   effects of magnetic fields and charged plates 223–4
   harmful effects 220
   uses of 226
Big Bang theory 255–6
black dwarf stars 251, 252
black holes 252
boiling 175, 176
   comparison with evaporation 177–8
boiling points 175
Boyle's Law 181–3
braking distance 34–6
brightness of stars 253

cars
   re-fuelling 90
   safety features 38–9
   stopping distances 34–6
Celsius temperature scale 180
centre of gravity 27–8
Ceres 242
chain reaction, nuclear fission 233
changes of state 175–8
charge (Q) 62–5, 84
   effects on radioactive emissions 223–4
   forces between charges 87
   see also electrostatic charge
chemical energy 123, 124, 140
circuit breakers 59
circuits see electric circuits
collisions 41–3
colours of light 111–12
comets 243
compressions, sound waves 112–13
condensation 175
conduction of thermal energy 127, 129, 132
conductors, electrical 58, 84–5
conservation of energy 123
conservation of momentum 41–3
contact forces 20
control rods, nuclear reactors 234
control systems 81
convection 128, 129, 132, 134
conventional current 63
coolant, nuclear reactors 234
cooling times 132
cosmic microwave background (CMB) radiation 256
cosmic rays 220
cosmology
   Big Bang theory 255–6
   Doppler red-shift 257–60
   Hubble's Law 261
coulombs (C) 62
critical angle 109–10
crumple zones 39
current-carrying wires, magnetic fields around 197–8
current–voltage graphs 80, 81
cycle helmets 39

d.c. motor 203–4
deceleration 6, 8–9
density 154–5
   of an irregularly shaped object 159–60
   of a liquid 157–8
   optical 105
   of a regularly shaped object 156–7
diodes 80–1, 83
direct current (d.c.) 54
   d.c. motor 203–4
direct proportion 45
dispersion of light 112
dissipation of energy 144, 172–3
   calculation of 173–4
distance moved 8, 13

calculation from velocity–time graph 10, 12
relationship to average speed and time 17, 18–19
distances 7
   in space 237
distance–time graphs 6–7
domains of magnetic materials 196
Doppler effect 115–16
   red-shift 256–60
double glazing 133
double insulated devices 60
drag 29
drawing pins 163
dwarf planets 242
dynamos 209

Earth 239, 242
   orbital speed 245, 246
earth wires 60
earthing 84, 90
echoes
   calculating the speed of sound 119–20
   sonar systems 120–1
efficiency 125–6, 145
   of transformers 214–15
elastic (strain) energy 123
elastic behaviour 45
elastic objects, Hooke's Law 43–6
electric charge see charge
electric circuit symbols 66
electric circuits 53
   ammeters 69
   parallel 68, 72–3
   series 67, 72
   voltmeters 70
   water pipe comparison 71
electric current (I) 53, 62–3
   conventional current 63
electrical components, power of 55–7
electrical devices 53–4
   power rating 54
electricity
   mains supply 53
   potential hazards 58
   safety measures 58–60
   transmission of 215–16
   units of 53
electricity generation
   comparison of methods 151–3
   from fossil fuels 148
   hydroelectric power 144
   nuclear power 148, 232–4
   from renewable resources 148–50
   use of electromagnetic induction 209
electromagnetic induction 207–8
   in a coil 208–9
electromagnetic spectrum 98
electromagnetic waves 93, 98
   dangers of 99
   summary of 101
   uses of 99–100
electromagnets 196–7, 200–1

in loudspeakers 205
'electron guns' 206
electrons 84, 85, 217, 218
    and charge 86
electrostatic charge 84
    dangers of 90
    detection of 88
    effects of 89
    production of 86
    uses of 90
    see also charge
electrostatic energy 124
elements 218
endoscopes 111
energy
    conservation of 123
    gravitational potential 138–43
    kinetic 136–7
    requirements for changes of state 176
    thermal 169
    useful 125
energy dissipation 144, 172–3
    calculation of 173–4
energy resources 147
    nuclear power 233–5
energy stores 123–4
energy transfers 53–4, 55–7, 124
    associated units 123
    domestic radiators 134
    efficiency 125–6
    examples of 124–5
    and gravitational potential energy 138–43
    multi-stage 144
    relationship to charge and voltage 64–5
    relationship to power 145
    Sankey diagrams 126–7
    thermal energy 127–31
    thermal energy loss reduction 133, 134
equations of motion 16–19
equilibrium 49–51
errors, parallax 2–3, 157
evaporation 177–8

falling objects 31–3
    determining speed from height dropped 141–2
    energy transfers 138–40
fibre optic cables 110–11
fire extinguishers 189
fission, nuclear 147, 232–3
    nuclear reactors 233–5
Fleming's left-hand rule 202–3, 224
floor insulation 133
forces 20–1
    between charges 87
    extension of a spring 45–6
    friction 29–30
    gravitational 239–40
    magnetic 192
    moments 46–52
    Newton's Laws of Motion 22–4, 37–8
    relationship to pressure 161–2
    units of 2
    weight 25–6
    and work done 135–6
fossil fuels 147
    advantages of 151
    disadvantages of 152
    use in power stations 148
free fall, acceleration of 31, 138
frequency of a wave 92
    sound waves 114, 121, 122
friction 24, 29–30
Fukushima nuclear power station 225
fuses 58–61
fusion, latent heat of 176
fusion, nuclear 147, 235

galaxies 236–7
    Milky Way 238
    red-shift 256–60
    velocity of 259–60
Galileo 31
gamma (γ) radiation 98, 101, 219, 228
    dangers of 99
    effects of magnetic fields and charged plates 223–4
    harmful effects 220
    uses of 100, 226, 227
gases
    Boyle's Law 181–3
    density of 155
    Pressure Law 184–6
    pressure of 179–80
    properties of 167–8
    relationship between temperature and volume 186–8
Geiger–Müller tube 221
generators 210–11
geostationary satellites 248
geothermal energy 150
    disadvantages of 153
glass, refractive index of 108–9
global warming 152
gold leaf electroscope 88
gradients 7, 10
graphs
    current–voltage 80, 81
    distance–time 6–7
    velocity–time 8–12, 19
gravitational field strength 25–7, 31, 138, 239–40
    units of 2
gravitational potential energy 123, 124, 138–43
    orbiting objects 241

half-life 229–32
Halley's comet 243
hard magnetic materials 196
heating effect of electricity 54, 59
helium shell burning, white dwarf stars 251
hertz (Hz) 91, 92
Hertzsprung–Russell (HR) diagram 254
homes, thermal insulation 133
Hooke's Law 43–6
horseshoe magnets 195
Hubble, Edwin 260
Hubble's Law 261
hydroelectricity 144, 150
    disadvantages of 153
hydrogen core burning, main sequence stars 250

hydrogen shell burning, red giant stars 251

ideal gases 179
images in a plane mirror 103–4
induced voltage 207–8
    in a coil 208–9
infrared radiation 98, 101, 128, 130–2
    dangers of 99
    uses of 100
infrasound 114
inkjet printers 90
instantaneous speed 7
insulation, thermal 133, 172–3
insulators, electrical 58, 84–5
internal energy 123
ionising radiation 220
    detection of 221–3
irradiation 225–6
isotopes 218

joules (J) 123
Jupiter 239, 242
    gravitational field strength 239–40
    orbital speed 245

Kelvin scale 180
kilograms (kg) 2, 25
kinetic energy 123, 124–5, 136–7
    of falling objects 138–43
    of gases 179–80
    orbiting objects 241
    and temperature 180
kinetic model of matter 166–8

lamps
    resistance of 79
    in series and parallel circuits 72–3
latent heat of fusion 176
latent heat of vaporisation 176
length
    measurement of 2–3
    units of 2
Leslie's cube 130–1
light, visible 98, 101
    colours of 111–12
    Doppler effect 116
    properties of 102
    reflection of 102–4
    refraction of 105–9
light gates 5
light rays 102
light years 237
light-dependent resistors (LDRs) 83
light-emitting diodes (LEDs) 83
liquids
    density of 157–9
    pressure in 164–6
    properties of 167, 168
live wires 60
loft insulation 133
longitudinal waves 92–3
    sound 112–14
loudspeakers 82, 205

magnadur magnets 194
magnetic energy 124
magnetic field lines 192
    plotting using a compass 193

magnetic fields 192-3
   around a current-carrying wire 197–8
   around a solenoid 199
   effects on radioactive emissions 223–4
   force on a current-carrying conductor 201–3
   forces on charged particles 206
   interaction of 193–4
   uniform 194–5
magnetic materials 191
   hard and soft 196–7
magnetism 191
   electromagnets 196–7, 200–1, 205
   testing for 195–6
main sequence stars 250
   Hertzsprung–Russell (HR) diagram 254
Mars 239, 242
   gravitational field strength 239–40
   orbital speed 245
mass
   and kinetic energy 136–7
   measurement of 4
   and momentum 36–7
   Newton's Laws of Motion 22–3
   relationship to weight 25–7
   units of 2
mass number (nucleon number) 217
measurement
   of length 2–3
   of mass 4
   of time 4–5
measuring cylinders 157
melting 175, 176
melting points 175
meniscus 157
Mercury 239, 242
   orbital speed 245
metals, conduction of electricity 84
metres (m) 2, 3
microphones 82
microwaves 98, 101
   dangers of 99
   uses of 99
Milky Way 237–8
mirrors 102–4
moderator, nuclear reactors 233–4
moments 46–7
   principle of 48–52
   units of 2
momentum 36–7
   conservation of 41–3
   and Newton's second law 37–8, 40
   safety aspects 38–9
   units of 2
Moon
   gravitational field strength 240
   orbital speed 247
moons 240, 244
motion
   distance–time graphs 6–7
   equations of 16–19
   investigation of 14–15
   Newton's Laws 22–4, 37–8
   units of 2
   velocity–time graphs 8–12, 19
motors, d.c. 203–4
musical instruments 114

natural satellites 244
nebulae 249
Neptune 239, 242
   orbital speed 245
neutral point, interacting magnetic fields 194
neutral wires 60
neutron emission 228
neutron stars 252
neutrons 85, 217, 218
newtons (N) 2
Newton's Laws of Motion 22–4
   and momentum 37–8
non-renewable energy sources 147
   advantages of 151
   disadvantages of 152
normal 102
nuclear energy 124
nuclear fission 147, 232–5
nuclear fusion 147, 235
   in main sequence stars 250
   in red giant stars 251
   in red supergiant stars 252
nuclear power stations 148, 225
   disadvantages of 152
   nuclear reactors 233–5
nucleon number (A) 217
nucleus of an atom 85, 217
nuclide notation 217–18

ohms (Ω) 64
Ohm's law 74–7
optical density 105
optical devices 100
optical fibres 110–11
orbital radius 246
orbital speed 245–8
orbits 21, 240–1
   of comets 243
   of planets 241–2
   of satellites 244
oscilloscopes 122

paint guns, electrostatic 90
parachutes 32–3
parallax error 2–3, 157
parallel circuits 68
   advantages of 73
   with identical lamps 72–3
pascals (Pa) 161
pendulums
   dissipation of energy 144
   period of 4
periscopes 111
photocopiers 90
photovoltaic (PV) cells 148–9
   disadvantages of 152
pitch 114, 121
pivots 46
plane mirrors 102–4
plane waves 93–4
planets 238–9
   dwarf 242
   gravitational field strength 239–40
   orbital speeds 245–6
   orbits 240–2
plugs, electric 59–60
Pluto 242

poles of magnets 191
power 145–6
   definition of 55
   of electrical components 55–7
power 'loss', in cables 216
power rating, electrical devices 54
power stations 148
pressure 161
   atmospheric 164
   in liquids 164–6
   practical examples of 163
Pressure Law 184–6
pressure of gases 179–80
   relationship to temperature 184–6
   relationship to volume 181–3
principle of moments 48–52
prisms 111
proportionality, limit of 45–6
proton number (Z) 217
protons 85, 217, 218
protostars 249

radiation 124
   background radiation 220
   infrared 128, 130–2
   ionising 220
radiators 134
radio waves 98, 101
   uses of 99
radioactive carbon dating 227
radioactive contamination 225
radioactive decay 219, 227–8
   half-life 229–32
radioactive emissions 219
   detection of 221–3
   effects of magnetic fields and charged plates 223–4
   harmful effects 220
   safety measures 221
radioactive isotopes 218
   uses of 226–7
radioactive waste 225
radiotherapy 227
radon gas 220
rarefactions, sound waves 113
reaction times 4, 34
reactions, Newton's 3rd Law 24
rectification 80–1
red giant stars 250–1
   Hertzsprung–Russell (HR) diagram 254
red-shift 116, 256–60
red supergiant stars 252
   Hertzsprung–Russell (HR) diagram 254
reflection 94, 102–4
   see also total internal reflection
refraction 94–6, 105–9
refractive index 106–7
   calculation of 108–9
   relationship to critical angle 110
relays 200–1
renewable energy sources 147
   advantages of 152
   disadvantages of 152–3
   electricity generation 148–50
   solar heating 151
resistance (R) 54, 64
   of a diode 80–1
   of a filament lamp 79

of an unknown resistor 78–9
water pipe comparison 71
resistors, variable 82
resultant force 20, 21
Newton's Laws of Motion 22–3
right hand grip rule 198
ripple tanks 93
reflection 94
refraction 94–5
rockets 24

Sankey diagrams 126–7
satellites 244
geostationary 248
orbital speed 248
Saturn 239, 242
gravitational field strength 239–40
orbital speed 245
scalar quantities 20, 21–2
seat belts 39
seconds (s) 2
series circuits 67, 72
short circuits 58
SI (Système International) system 3
Snell's law 106–7
snow shoes 163
soft magnetic materials 196
solar cells 148–9
disadvantages of 152
solar heating 151
solar system 238–9
asteroid belt 242
comets 243
dwarf planets 242
orbits 240–2
solenoids 199
solidification 175, 176
solids, properties of 166–7, 168
sonar systems 114, 120–1
sound waves 93
characteristics of 121
Doppler effect 115
echoes 119–20
frequency of 114
oscilloscope traces 122
production of 113–14
properties of 112–13
sonar systems 120–1
speed of 116–19
specific heat capacity 169–70
determination of 171–2
spectra, of visible light 112
speed 21
calculation from distance–time graphs 7
of electromagnetic waves 98
of falling objects 141–2
and kinetic energy 136–7
of light 102
measurement of 14–15
orbital 245–8
of sound 112, 116–19
units of 2
of a wave 92
springs, Hooke's Law 43–6
stars
absorption spectra 257–8
brightness of 253

classification by colour 249
Hertzsprung–Russell (HR) diagram 254
lifecycle of 249–52
main sequence 250
nuclear fusion 147
red giant 250–1
red supergiant 252
white dwarf 251
states of matter 166–9
changes of state 175–8
stellar evolution 249–52
step-up and step-down transformers 213
use in electricity transmission 216
stopping distance of a car 34–6
stopwatches 5
streamlining 29
Sun 238, 249
brightness of 253
gravitational field strength 240
nuclear fusion in 235
sunlight exposure, dangers of 99
supernovae 249, 252

temperature 180
and changes of state 175–8
temperature of a gas
relationship to pressure 179–80, 184–6
relationship to volume 186–8
terminal velocity 32–3
thermal energy 123, 124–5
dissipation of 172–4
specific heat capacity 169–72
thermal energy transfer 127–8
consequences of 132
experiments 129–31
thermal insulation 133
vacuum flasks 134
thermistors 82–3
thinking distance 34–6
ticker tape timers 14–15
tidal energy 150
disadvantages of 153
time
measurement of 4–5
units of 2, 7
time period of a wave 92
total internal reflection 109–10
uses of 110–11
transducers 82
transformers 212, 214
efficiency of 214–15
step-up and step-down 213
use in electricity transmission 216
transverse waves 92, 93
turbines 209
see also wind turbines
turning effects (moments) 46–7
principle of moments 48–52

ultrasound 114
ultraviolet radiation 98, 101
dangers of 99
uses of 100
units, SI system 3
universe 236
Big Bang theory 255–6
expansion of 261
Uranus 239, 242

orbital speed 245
useful energy 125

vacuum flasks 134
Van de Graaff generator 89
vaporisation, latent heat of 176
vector quantities 20, 21–2
momentum 36–7
velocity 7, 21
and kinetic energy 136–7
terminal 32–3
units of 2
velocity–time graphs 8–12, 19
Venus 239, 242
gravitational field strength 239–40
orbital speed 245
virtual images 103
volmeters 70
voltage 55, 64–5
induced 207–9
volume of a mass of gas
relationship to pressure 181–3
relationship to temperature 186–8

wall insulation 133
watts (W) 145
wave energy 149
disadvantages of 153
wave equation 92
wavefronts 93
wavelength 91, 121
waves
associated terms 91–2
calculations involving 96–7
electromagnetic 93, 98–100
longitudinal 92–3
properties of 91
reflection 94
refraction 94–6
sound 112–14
transverse 93
units 91
see also light, visible
weight
centre of gravity 27–8
falling objects 32–3
relationship to mass 25–7
white dwarf stars 251
Hertzsprung–Russell (HR) diagram 254
white light, dispersion of 112
wind turbines 149
disadvantages of 152
work 135–6

x-rays 98, 101
dangers of 99
uses of 100